Lecture Notes in Computer Sc

Commenced Publication in 1973
Founding and Former Series Editors:
Gerhard Goos, Juris Hartmanis, and Jan van Leeuwen

Niki Trigoni   Andrew Markham
Sarfraz Nawaz (Eds.)

GeoSensor
Networks

Third International Conference, GSN 2009
Oxford, UK, July 13-14, 2009
Proceedings

 Springer

Volume Editors

Niki Trigoni
Andrew Markham
Sarfraz Nawaz
Oxford University Computing Laboratory
Wolfson Building, Parks Road, Oxford, OX1 3QD, UK
E-mail: {niki.trigoni,andrew.markham,sarfraz.nawaz}@comlab.ox.ac.uk

Library of Congress Control Number: 2009930106

CR Subject Classification (1998): C.3, J.2, C.2.4, C.2.1

LNCS Sublibrary: SL 3 – Information Systems and Application, incl. Internet/Web and HCI

ISSN 0302-9743

ISBN 978-3-642-02902-8 Springer Berlin Heidelberg New York

springer.com

© Springer-Verlag Berlin Heidelberg 2009

Typesetting: Camera-ready by author, data conversion by Scientific Publishing Services, Chennai, India
Printed on acid-free paper   SPIN: 12711901   06/3180   5 4 3 2 1 0

Preface

This volume serves as the conference proceedings for the Third International Conference on GeoSensor Networks (GSN 2009) that was held in Oxford, UK, in July 2009. The conference addressed issues related to the deployment of geosensor networks and the collection and management of real-time geospatial data. This volume includes papers covering a variety of topics, ranging from sensing, routing and in-network processing, to data modelling, analysis, and applications. It reflects the cross-disciplinary nature of geosensor networks by bringing together ideas from different fields, such as geographic information systems, distributed systems, wireless networks, distributed databases, and data mining.

The papers included in this volume push the frontiers of GSN research in several new directions. First, they shift the traditional paradigm of fixed resource-constrained sensor networks to more versatile scenarios involving mobile sensors with varying levels of computation, sensing, and communication capabilities. Second, emerging applications, such as urban sensing, are raising interesting new GSN challenges, such as participatory sensing, as well as the need to standardize protocols and interfaces, and integrate multiple sensor networks. Finally, technological advances in geosensor networks have enabled the collection of vast amounts of real-time high-resolution spatio-temporal data; the sheer volume of these data raises the need for efficient data modelling, analysis, and pattern extraction techniques.

In order to address these topics, the GSN 2009 conference brought together leading experts in these fields, and provided a 2-day forum to present papers and exchange ideas. The papers included in this volume are select publications presented during the GSN 2009 conference that went through a rigorous refereeing process; the volume also includes a small number of invited papers. More information about the scientific background of geosensor networks and an outline of the papers included in this volume may be found in the Introduction.

We greatly appreciate the many people who made this happen. Specifically, we would like to acknowledge the financial support of EOARD, the European Office of Aerospace Research and Development. We would also like to thank the University of Oxford for their administrative support, and for hosting various events of the conference. We would especially like to thank Elizabeth Walsh, Mike Field and Katie Dicks, as well as all the members of the Sensor Networks Group at the Oxford University Computing Laboratory, who helped in the organization of the GSN 2009 conference. Last, but not least, we would like to thank everybody who helped in the production of this volume, and, in particular, the authors and participants of the conference, the Programme Committee members, and Springer.

May 2009

Niki Trigoni
Andrew Markham
Sarfraz Nawaz

Organization

Organizers

Niki Trigoni   University of Oxford, UK
Andrew Markham   University of Oxford, UK
Sarfraz Nawaz   University of Oxford, UK

Programme Committee

Alastair Beresford   University of Cambridge, UK
Ioannis Chatzigiannakis   University of Patras, Greece
Alfredo Cuzzocrea   University of Calabria, Italy
Antonios Deligiannakis   Technical University of Crete, Greece
Matt Duckham   University of Melbourne, Australia
Dinos Ferentinos   University of Athens, Greece
Alvaro Fernandes   University of Manchester, UK
Sanjay Jha   University of New South Wales, Australia
Vana Kalogeraki   University of California, Riverside, USA
Yannis Kotidis   Athens University of Economics and Business, Greece
Antonio Krueger   University of Muenster, Germany
Lars Kulik   University of Melbourne, Australia
Andrew Markham   University of Oxford, UK
Kirk Martinez   University of Southampton, UK
Cecilia Mascolo   University of Cambridge, UK
Sarfraz Nawaz   University of Oxford, UK
Silvia Nittel   University of Maine, USA
Peter Pietzuch   Imperial College London, UK
Mirek Riedewald   Northeastern University, USA
Monika Sester   LUH, Germany
Egemen Tanin   University of Melbourne, Australia
Theodore Tsiligiridis   Agricultural University of Athens, Greece
Sonia Waharte   University of Oxford, UK
Peter Widmayer   ETH, Switzerland
Michael Worboys   University of Maine, USA
Eiko Yoneki   University of Cambridge, UK

Introduction to Advances in Geosensor Networks

Niki Trigoni, Andrew Markham, and Sarfraz Nawaz

Oxford University Computing Laboratory
Wolfson Building, Parks Road
Oxford OX1 3QD
Niki.Trigoni@comlab.ox.ac.uk
acmarkham@gmail.com
Sarfraz.Nawaz@comlab.ox.ac.uk

Over the last decade, significant advances in sensor and communication technology have pushed the frontiers of geosensor networks from the research lab to real deployments in a variety of novel applications. The main shift from traditional GIS research started with the ability to deploy untethered sensors in remote areas, and to collect data in real-time and at varying spatial and temporal scales and granularities. In order to realize these capabilities in an energy-efficient manner, ongoing research efforts have focused on various aspects of network functionality, ranging from sampling and topology control, to routing and in-network processing. Significant work is also directed to the modelling of geospatial events, and the offline analysis of data after they are collected.

As the field of geosensor networks becomes more mature, novel applications are continuously emerging, with different requirements and research challenges. For example, remote environmental sensing applications typically involve battery-powered nodes that cover large geographical areas, and pose interesting energy-efficiency and scalability problems. Urban-sensing applications typically involve a mixture of fixed and mobile sensor nodes with rechargeable batteries, and significant computational and storage capabilities. Such applications raise challenges of frequent network disconnections because of mobility, limited bandwidth resources, and interoperability between sensor devices and networks.

The papers collected in this volume cover a wide spectrum of key areas in geosensor network research, including sensing, routing, clustering, in-network compression, in-network filtering, query processing, data analysis, pattern extraction, modelling of geospatial events, and application requirements. These topics are clustered into four main sections. The first section on *Sensing and In-Network Processing* includes papers that investigate the areas of intelligent sampling, in-network compression, and in-network filtering and estimation. The second section on *Routing* introduces papers that explore routing and node placement techniques in geosensor networks. The third section on *Data Analysis and Modelling* focuses on data management issues: how to model high-level geospatial events from low-level sensor readings, and how to analyze and extract patterns from large volumes of geospatial data. Finally, the section on *Applications and Integration* introduces papers focusing on specific and general requirements of geosensor applications.

1   Sensing and In-Network Processing

The first section of this book is dedicated to papers that deal with compressive sampling and in-network processing in geosensor networks. Since sensor networks are often deployed in remote areas with limited energy resources to sample physical phenomena, efficient sampling algorithms and en-route compression strategies are required to prolong their lifetime.

Energy-efficient transfer of the large amount of data generated by sensor nodes to a single collection point is an important challenge in sensor networks. To this end, a number of in-network data compression schemes have been introduced that try to exploit the high correlation and redundancy in data collected at nearby nodes. Most of these techniques employ a wavelet or some other form of signal processing transform in a distributed manner. However, the performance of such a transform depends on the underlying routing structure over which data are forwarded toward the sink node. The paper by Tarrio, Valenzise, Shen, and Ortega addresses this issue when the routing tree over which the compression is performed is built in a distributed manner. The main result presented in the paper is that the tree structures used for routing are generally good enough to perform a compression transform and the trees that are specifically designed to perform compression do not deliver a significant performance gain.

Compressed sensing (also known as compressive sensing) has recently emerged as an alternative to in-network compression techniques for sensor networks. Instead of collecting a large amount of correlated data and compressing it on its way to the sink node, compressed sensing proposes the collection of a small number of measurements that can later be used to reconstruct the original signals at the sink node by applying numerical optimization techniques. Thus this technique not only generates smaller amounts of data but it also offloads the computationally intensive processing to the sink node. The paper by Lee, Pattem, Sathiamoorthy, Krishnamachari, and Ortega proposes lowering the cost of performing these measurements in sensor networks by dividing the network into clusters of adjacent nodes. They study the effect of different clustering schemes on the signal reconstruction performance and show that such clustering mechanisms result in significant power savings in compressed sensing.

The paper by Zoumboulakis and Roussos takes in-network processing to the next step and performs what can be termed as in-network estimation. They propose a distributed and localized algorithm for determining the position of a pollutant-emitting point source. They use a Kalman filter-based *predict and correct* pattern to home in on the pollutant source. The main advantage of this approach is that instead of routing a large amount of data to a centralized collection point through a resource constrained network, the decision making is performed in the network at points where the data are being generated.

Due to the data-centric nature of geosensor networks, the quality of the collected data is also an important aspect of sensor network applications. If the data collected from the sensor network contain a significant amount of outliers, this not only affects the fidelity of the data delivered to the application, but it also wastes network resources. Thus, these outliers should be detected and removed in the network before being delivered to the central data collection point. The paper by Zhang, Meratnia, and Havinga addresses this specific challenge. They propose using Hyperellipsoidal SVM, where each sensor node solves a linear program with locally sensed or training data as inputs to determine a threshold. This locally computed threshold is then exchanged with one-hop neighbours. Each node can then detect an outlier by comparing each measurement with both its own threshold and those of its neighbours. The authors analyze the performance of this outlier detection algorithm using both synthetic and actual data collected from a real sensor network deployment.

2 Routing

Fundamental to the operation of geosensor networks are aspects related to the efficient gathering of information from the field. Although some of these issues have been covered in the previous section on sensing and in-network processing, papers concentrating on the actual relaying process are presented in this section.

The placement of gateway nodes within a network of sensor nodes is an important factor for deployment. Efficient spacing of these high-powered nodes reduces the number of hops that data need to be sent over, which impacts network lifetime and congestion. Sachdev and Nygard discuss the optimal placement of gateway nodes by using a Genetic Algorithm as a search technique to efficiently place these high-powered relay nodes such that the distance (evaluated either in terms of the number hops or physical distance) between the relay and the nodes within its cluster is minimized. The number of clusters is specified as a design parameter and the GA is used to determine good locations for the fixed number of relay nodes. To demarcate the clusters, a Voronoi diagram is used. The GA is used to minimize both the overall distance between the relays and their child nodes and also to ensure fair load-balancing among the clusters. The authors present results for different numbers of clusters, nodes, and distance metrics and demonstrate that their method is able to efficiently place relay nodes.

The previous paper considered predeployment issues. However, during network operation, it is necessary to fairly balance network load among nodes to prevent premature node exhaustion. Schillings and Yang present an interesting and novel approach to maximizing the wireless sensor network lifetime using Game Theory. Game Theory treats nodes as players in a game, where a player's goal is to maximize its own utilization of valuable resources, in this case, energy and information. However, nodes cannot act in isolation, as selfish, locally optimal decisions could result in poor global performance. To this end, nodes are tasked not only with their own survival, but also with the maintenance of a path to the sink. Thus, critical nodes close to the sink are treated as a valuable resource that must be preserved in order to ensure the survival of the network as a whole, by preventing segmentation. The authors compare their approach with a number of existing routing protocols and demonstrate how their Game Theoretic approach can prolong network lifetime by rotating the selection of the next hop candidate for a message.

Sensor network researchers often advocate the need for cross-layer optimization between the routing and query processing layers. The idea is that better routing decisions can be made if the routing layer is aware of the expected traffic incurred by the query processing layer. The paper by Deligiannakis, Kotidis, Stoumpos, and Delis proposes energy-efficient algorithms for building aggregation trees that take into account the expected participation of nodes in event monitoring queries. Recent values of node participation frequencies are used to estimate attachment costs to candidate parents. By choosing the parent with the lowest attachment cost, nodes are able to design significantly better aggregation trees than existing techniques. The presented algorithms are based on the aggregation and transmission of a small set of statistics in a localized manner. They can be used to minimize important metrics, such as the number of messages exchanged or the energy consumption in the network.

Turning to higher level IP-based networks, the paper by Gaied and Youssef shows how geographical information can be exploited in order to optimize the performance of a peer-to-peer overlay network for information dissemination. The advantage of a peer-to-peer network is in its distributed nature – there is no central controller responsible for routing. An overlay network abstracts the underlying physical network. In the original version of Chord (a standard peer-to-peer protocol) the mapping between a node's IP address and its hash key, which identifies it in the overlay, is random. This means that nodes which are physically close to one another are not necessarily neighbours in the overlay network. This results in inefficient routing, as message latency is very high. The authors show how, by using the coordinates of the nodes in the networks to augment the hashing function, a more efficient overlay network can be constructed. Approximate coordinates of nodes are inferred by using the Global Network Positioning (GNP) architecture, which determines location from round-trip-times to servers with known co-ordinates. Thus, this paper shows how geospatial information can be used as an effective means to improve the transfer of information through a network.

3   Data Analysis and Modelling

Geosensor networks have the ability to generate vast amounts of raw data at high spatial and temporal resolutions. However, once the data have been collected, there is a need to analyze it, to extract meaningful patterns that can be used to understand and model the underlying phenomenon. This section presents some interesting papers which tackle this problem, generally by determining what is required by the end user of the data. However, some of these techniques have the secondary, beneficial effect of reducing network load by taming data volumes.

RFIDs are becoming a ubiquitous part of daily life, with applications ranging from automatic toll collection on highways to prepaid subway fares. They are also used for supply chain monitoring and increasingly for wildlife monitoring. However, a problem with RFID technology is the sheer volume of data which is generated, making data mining and analysis difficult. Bleco and Kotidis present a number of algorithms that exploit spatial and temporal correlations in order to reduce the size of the raw RFID data stream. Their algorithms run as 'Edgeware' on multiple readers in order to reduce the volume of data which needs to be sent over the RFID based network. Their algorithms work by packing the data into tuples which act as aggregates. However, unlike prior work, they account for the fact that RFID readers have a high drop rate by including a percentage of data received in each tuple, in essence constructing a histogram of detection probability at each reader. This approach thus aggregates temporal patterns of RFID data received by a single reader. Following this, spatial patterns are also aggregated by identifying groups which consistently move together past multiple readers – for example, a pallet load of RFID-tagged products will not move independently but as a group that can be replaced by a surrogate ID. This further helps to reduce the volume of data. Using their algorithms (of which, both lossy and lossless variants are presented), they demonstrate that typical RFID data streams can be reduced by a factor of 4 with modest computational overhead.

In a similar thread to the prior paper, Masciari presents a method for trajectory clustering of location-based data (such as GPS logs or RFID traces). Trajectory clustering is necessary to efficiently analyze these vast volumes of data, by extracting common patterns between trajectories. To do so, the data is first discretized and translated into symbols. Principal Component Analysis (PCA) is used to identify frequently visited regions, and regions which are infrequently traversed are discarded from further analysis. Thus, PCA is used to reduce the size of the symbol alphabet, while still preserving the salient features of the trajectories. Once the trajectories have been translated from raw coordinates into an ordered list of symbols, similarities between trajectories are computed using a string editing metric. A number of examples from real-world datasets are presented which demonstrate how the PCA-based clustering method can identify similarities in data trajectories, thus simplifying further analysis.

When monitoring an environmental phenomenon (such as a pollution plume or wildfire), often what is needed for analysis is a thresholded or Boolean view of the sensor field – for example, a region is either on fire or it is not. This can be simplified by modelling the outlines or topologies of these regions. Jiang and Worboys have taken this a step further and present in their paper a theoretical overview of how to track changes in these topological regions over time. Following the theoretical elucidation, they then consider how to configure a sensor network in order to accurately capture these changes. Essentially, the goal is for the output of the sensor network to match the topological map of the measured phenomenon. However, real-world factors such as sensor density and inadequate communication ranges will result in inaccuracies in the overall representation. This is because the spatially continuous field is sampled at discrete intervals by sensor nodes which are unaware of their precise location. Thus, the authors stipulate constraints (both in terms of density and communication range) that need to be imposed on the node layout and configuration such that they correctly capture the topological changes.

4 Applications and Integration

Recent technological advances in sensors and communications are allowing the deployment of geosensor networks in a large class of applications, ranging from remote environmental monitoring (precision agriculture, ocean monitoring, early warning systems, and wildlife tracking) to urban-scale sensing (air quality monitoring, mapping of urban texture, and traffic monitoring). In these applications, geosensor networks are enabling real-time monitoring of interesting events with unparalleled spatial and temporal resolution.

Of particular interest are deployments of geosensor networks in large and often remote areas to warn of major environmental disasters, such as forest fires, river floods, volcanic activity, and seismic activity. These applications typically require real-time anomaly detection and instant warning reports, in order to allow rescue teams to take appropriate countermeasures and contain the extent of environmental damage and human losses. The paper by Santos (J.), Santos (R.), and Orozco provides a feasibility study for this type of application, and concludes that the sheer number of sensors necessary to cover large areas is prohibitively expensive in terms of deployment and maintenance costs. They propose the deployment of sparser networks, which do not provide full coverage of the area of interest, but are still capable of providing an early notice that the environmental event is in its beginning phase. They show that this second-best alternative is feasible from both a technical and an economic point of view.

Geosensor networks are allowing us to monitor remote and harsh regions at a scale that had not been possible before. This is not only opening up new frontiers of knowledge but also setting new challenges for us to solve. The paper by Martinez, Hart, and Ong describes details of a sensor network that was deployed at a glacier in Iceland to study its dynamics. Their network has generated environmental data that had not existed before, and has thus advanced our understanding of not only of the environment but also of deployment issues in harsh environments. The authors describe the design of their very low power sensor probes and base-station, and also outline their experiences from the actual deployment.

In terms of urban-scale sensing, a novel paradigm for data acquisition, has recently emerged often referred to as participatory sensing. The idea is to move away from the traditional paradigm of fixed special-purpose network infrastructures, and to exploit the spontaneous movement of people holding wireless devices with sensor capabilities. For example, the paper by Shirami-Mehr, Banaei-Kashani, and Shahabi considers the scenario of collaborative collection of urban texture information by a group of participants with camera-equipped mobile phones. Because of the limited user participation time, it is critical to select a minimum number of points in the urban environment for texture collection. The authors prove that this problem is NP-hard and propose a scalable and efficient heuristic for viewpoint selection. They also highlight the challenge of assigning viewpoints to users (taking into account their preferences and constraints) as an interesting direction for future work.

Given the large variety of existing and future geosensor applications, it is essential to understand the main driving factors for the deployment of real sensor systems. The paper by Kooistra, Thressler, and Bregt presents an expert survey on the user requirements and future expectations for sensor networks. The survey investigates the role of several factors in sensor network development: (1) the need for real-time data at varying spatio-temporal scales; (2) standardized protocols, interfaces, and services; (3) advances in sensor and communication technology; (4) broad user community and range of applications; (5) community participation as data providers; and (6) increasing openness in data policies. The authors conclude that technology and applications are the main driving factors for sensor network deployment. However, standardization, privacy, and data quality can significantly facilitate further development, and encourage community participation both as data sources and data providers. The survey respondents also highlight the need to create profitable business models from future geosensor network applications.

A practical mechanism to increase the value of existing geosensor network systems is to allow their integrated use by emerging pervasive applications, such as crisis management. The incorporation and correlation of data coming from various networks is an interesting research problem which is explored in the paper by Casola, Gaglione, and Mazzeo. The authors propose an integration framework based on a wrapper-mediator architecture: The mediator is responsible for forwarding user requests to different networks, while the wrappers are responsible for translating the incoming queries and forwarding them to the underlying sensors. The integration of multiple geosensor networks is a multi-faceted problem with many promising future directions in the areas of data quality assurance, resource sharing by multiple applications, and secure and privacy-preserving data sharing.

Another approach to facilitate application development is to develop high-level abstractions that view a sensor network as a system which can be tasked to execute high-level tasks without the specification of low-level details by the network user. This is in contrast to most recent work which has focused on distributed algorithms that are implemented at the node level and are targetted at a specific application. The paper by Stasch, Janowicz, Broring, Reis, and Kuhn proposes an ontological view of the sensors and sensor network, and discusses a stimulus-centric algebraic framework. The authors propose developing a programming API that could be used to abstract away the complex task of programming and development faced by current geosensor users.

5   Outlook and Open Issues

The collection of papers in this volume demonstrates the breadth and scope of research in the growing field of geosensor networks. In part due to technological advances, the adoption of geosensor networks is becoming more and more widespread. Devices are being deployed in extreme and remote areas, from the bottom of the sea to deep within glaciers. Coupled with improvements in technology is an increase in public awareness of issues such as climate change and carbon capture. From a forecasting perspective, there is a need for monitoring systems that are able to predict environmental disasters such as hurricanes and tsunamis. Thus, the demand for widespread data collection is likely to push the development and deployment of geosensor networks further. Although this volume has presented novel research in this arena, there are still many directions that this broad field can take in the future.

One such area of research is the incorporation of mobility into the sensing process. The majority of work to date concentrates on static sensor networks. Greater spatial resolution of sensed data can be obtained by moving sensor nodes over a region of interest. Thus, a fusion between the (currently) disparate fields of robotics and geosensor networks is likely to occur to a greater degree than at present. Unmanned vehicles can be directed to points of interest and collaboratively work together to achieve a certain sensing goal, such as locating and investigating the precise location of a chemical contaminant. These unmanned vehicles can be terrestrial, aquatic, or aerial, and as such, are likely to have different environmental constraints, such as ocean currents or reduced radio communication range due to buildings. Thus, a large amount of research is needed to formulate exploration strategies for these different scenarios that are both communication efficient and distributed. Another aspect is that of passive mobility, where sensors carried by people or animals act as random sampling agents over a sensor field. This is where the sheer volume of observations that are made (for example, by using mobile phone cameras) compensates for the lack of control over mobility. Lastly, most existing geosensor networks assume that users are passive observers of events. In certain classes of applications, such as disaster management, sensor observations influence user decisions, and consequent user actions will impact the evolution of monitored events. These complex feedback systems raise challenging issues worthy of future investigation.

As geosensor network applications continue to become a part of our daily lives, the issues of privacy and security become critical. In an owner-managed sensor network, security is relatively easy to ensure. However, when multiple sensor-enabled devices (which are not centrally owned or controlled) are used to gather data, security is a massive issue, especially if devices can be retasked to alter the sensors used and the sensing behavior. Such modification of behavior can have a variety of effects ranging from impacts on battery lifetime to breaches of privacy by relaying private conversations or images. Without trust that the data are not going to be misused and privacy compromised, the adoption by the public of participatory networks will be slow and limited.

Lastly, there is the problem of interoperability and standardization and related to this, issues of data ownership and sharing. There is a need for the simple exchange of geospatial information between multiple entities. In addition, there is also a strong case for sharing or selling collected sensor data. This further highlights the need for common formats that allow data from multiple sources to be combined. This will allow more detailed models of complex phenomena to be determined, by reusing data possibly collected for an entirely different purpose.

The collection of papers in this volume has provided a cross-section of the state of the art in the field of geosensor networks. In this section, we have outlined a few additional research topics in geosensor networks. It is our hope that it will serve as a useful reference to researchers in this growing, cross-disciplinary field, and we welcome participation in future GSN conferences.

Table of Contents

Applications and Interoperability

Distributed Network Configuration for Wavelet-Based Compression in Sensor Networks

Paula Tarrío[1,*], Giuseppe Valenzise[2], Godwin Shen[3], and Antonio Ortega[3]

[1] Departamento de Señales, Sistemas y Radiocomunicaciones,
Universidad Politécnica de Madrid, Madrid, Spain
[2] Dipartimento di Elettronica e Informazione,
Politecnico di Milano, Milan, Italy
[3] Department of Electrical Engineering,
University of Southern California, Los Angeles, California, USA
paula@grpss.ssr.upm.es, valenzise@elet.polimi.it,
godwinsh@usc.edu, ortega@sipi.usc.edu

Abstract. En-route data compression is fundamental to reduce the power consumed for data gathering in sensor networks. Typical in-network compression schemes involve the distributed computation of some decorrelating transform on the data; the structure along which the transform is computed influences both coding performance and transmission cost of the computed coefficients, and has been widely explored in the literature. However, few works have studied this interaction in the practical case when the routing configuration of the network is also built in a *distributed* manner. In this paper we aim at expanding this understanding by specifically considering the impact of distributed routing tree initialization algorithms on coding and transmission costs, when a tree-based wavelet lifting transform is adopted. We propose a simple modification to the collection tree protocol (CTP) which can be tuned to account for a vast range of spatial correlations. In terms of costs and coding efficiency, our methods do not improve the performance of more sophisticated routing trees such as the shortest path tree, but they entail an easier manageability in case of node reconfigurations and update.

Keywords: In-network compression, wavelet lifting, distributed routing algorithms, collection tree protocol, shortest path tree.

1 Introduction

Data gathering in Wireless Sensor Networks (WSN) is generally deemed to be energy-consuming, especially when the density of sensor deployment increases, since the amount of data to be transmitted across nodes grows as well. The observation that, for many naturally occurring phenomena, data acquired at

* This work has been partially sponsored by the Spanish Ministry of Science and Innovation, under grant BES-2006-13954.

N. Trigoni, A. Markham, and S. Nawaz (Eds.): GSN 2009, LNCS 5659, pp. 1–10, 2009.
© Springer-Verlag Berlin Heidelberg 2009

(spatially) neighboring nodes are correlated has suggested the use of en-route compression techniques, which directly send the acquired data to the sink in a compressed form. Such in-network compression methods and their interaction with routing have been extensively studied in the literature, both from a theoretical point of view [1,2] and in practical scenarios, using Slepian-Wolf coding [3], opportunistic compression along the shortest path tree [4] or wavelet transforms [5,6,7]. Nevertheless, only limited work [8] has been done on understanding the problems associated with distributed node configurations, i.e., how to build in a distributed fashion a routing structure along which to compress and gather data. This appears to be particularly appealing when the number of inter-node (or node-to-sink) messages is to be minimized in order to reduce initialization or reconfiguration times (e.g., in dynamic networks with mobile nodes). In the SenZip architecture [8], this aspect has been considered for the case of a tree-based spatial decorrelating transform, where the tree is built in a distributed way allowing only local message exchanges between nodes. However, a clear understanding of how different distributed routing initialization schemes impact the performance of a distributed transform is still missing.

This work builds on the distributed approach proposed in SenZip in order to analyze the effects of distributed tree-building algorithms on coding and transmission costs. As in SenZip, we focus on the in-network data compression scheme described in [7]. The key point of this algorithm is the computation of a 2D wavelet transform using a lifting scheme, where the lifting is operated along an arbitrary routing tree connecting nodes irregularly spaced in a sensor network. In SenZip, the collection tree protocol (CTP) [9], a tree-based collection service included in TinyOS, is employed to build this tree incrementally. We extend this CTP algorithm in order to fit naturally to a lifting scheme; moreover, we simplify tree management in the case of possible node reconfigurations. The proposed approach is compared with two distributed versions of the shortest path tree routing, which achieve better performance but with a higher reconfiguration complexity. We also compare the costs of transmission when the wavelet transform compression is employed with those incurred in the simple raw data gathering case. We observe that, even in the distributed setting, the gain of using a transform to decorrelate data is significant with respect to just forwarding data from nodes to the sink. Furthermore, we show that the gain of the transform increases when the average depth of peripheral nodes (leaves) of the routing tree with respect to the sink gets larger, i.e., when the average hop length in the tree is smaller (for a fixed density of the nodes in the network), as this leads to more highly correlated data (assuming that data in two nodes tends to be more correlated if the two nodes are closer to each other). Our study suggests that trees that are in general good for data gathering, will also be good when a transform is used, while trees specifically designed to improve transform performance will reduce only minimally the total costs, or will worsen them due to routing sub-optimality.

The rest of this paper is organized as follows. In Section 2, we provide a brief survey of previous work on en-route data compression, including the 2D wavelet transform of [7] used in our experiments. In Section 3 we present a tree-based collection algorithm to construct the routing tree for compression. In Section 4 we use simulations to study the impact of different distributed tree-building strategies on transmission costs. Our conclusions are in Section 5.

2   Related work

In-network data compression techniques include, among others, the distributed KLT proposed in [10], wavelet based methods [11,6,5,12,7], networked Slepian-Wolf coding [3], and, recently, distributed compressive sensing [13]. In particular we focus on wavelet transforms based on lifting, which can easily be used even in the case of arbitrary node positions. While early approaches required backward data transmission [12] (i.e., away from the sink), these bidirectional data flows (i.e. from and towards the sink) lead to higher transmission costs. Instead unidirectional transforms, where data only flows towards the sink, are preferable [11,14,7]. We focus on the approach in [7], where a unidirectional way along any routing tree, e.g., shortest path tree (SPT), can be computed. The SPT guarantees the best path from a given node to the sink from the routing perspective, but obviously does not ensure that consecutive nodes in a path contain highly correlated data. Conversely a tree that seeks to minimize inter-node distance (e.g., a minimum spanning tree, MST), in order to lower the bitrate, may have a higher transport cost. In [15], this problem is addressed by proposing a joint optimization of routing and compression. A broader perspective is adopted in [16], where each node can collect data broadcasted by its neighbors (i.e. nodes that are spatially close), even in the case they are not directly connected in the tree structure. This enables the use of a routing-optimal tree (SPT) while leveraging the context information at the same time. Very recently, Pattem et al. [8] have demonstrated for the first time an architecture for en-route compression, based on the tree wavelet transform of [7], in which the routing (compression) tree is initialized in a *distributed* way, using CTP. In this paper, our goal is to evaluate different network routing initialization techniques in terms of their impact on overall costs for a given data representation quality.

In order to compress the data gathered by the sensor nodes in a distributed manner we use the 2D unidirectional lifting transforms proposed in [7,16]. These transforms are critically sampled and computable in a unidirectional manner, so that the transmission costs are reduced. Consider a sensor network with N sensor nodes and one sink, where x_n is the data gathered by node n. Let T be the routing tree, with the root of the tree corresponding to the sink (node $N + 1$). Let depth(n) be the number of hops from n to the sink in T, with depth$(N + 1) = 0$. To perform a lifting transform on T, we first split nodes into even and odd sets. This is done by assigning nodes of odd (resp. even) depth as odd (resp. even) in the transform. Data at odd nodes is then predicted using data from even neighbors on T, yielding detail coefficients. Then, even node data is updated using detail coefficients from odd neighbors on T. This can be done over multiple levels of decomposition, either directly on T or on other trees using the notion of "delayed processing" introduced in [16].

3   Distributed Tree Construction Algorithms

We outline in this section three distributed algorithms to initialize the routing tree on which the tree-based wavelet transform will be computed.

3.1   Modified Collection Tree Protocol (M-CTP)

The collection tree protocol (CTP) [9] is a tree-based data gathering approach that provides best-effort anycast datagram communication to the sinks in a multihop network. The protocol has an initialization phase in which one or more nodes advertise themselves as sinks (or tree roots) and the rest of the nodes form routing trees to these roots using the expected transmissions (ETX) as the routing gradient. This metric gives an estimate of the number of transmissions it takes for a node to send a unicast packet whose acknowledgement is successfully received. The ETX of a node is the ETX of its parent plus the ETX of the link to its parent and given a choice of valid routes, CTP chooses the one with the lowest ETX value. CTP assumes that the data link layer provides synchronous acknowledgements for unicast packets, so when the acknowledgement of a packet is not received, the packet is retransmitted.

In this work we have used a similar collection protocol, but instead of using the ETX metric, the protocol chooses the links with lowest inter-node distance. In the following, we refer to this modified version of the collection tree protocol as M-CTP. In the initialization phase, the sink broadcasts a 'Hello' packet using a given transmission power. All the nodes within its communication range receive this packet, and label themselves as 1-hop nodes. Then, using the same transmission power as the sink, each 1-hop node broadcasts another 'Hello' packet, which will be received by their neighboring nodes. Among these nodes, those which have not received previously any 'Hello' packet will become 2-hop nodes and they will continue with the initialization procedure in the same way. At the end, every node in the network will know their number of hops to the sink, from which it will be possible to determine the node parity (even or odd), and therefore the role of the node in the tree-based transform[7]. Furthermore, when a node receives a packet from another node, it measures the Received Signal Strength (RSS) of the message and uses this information to estimate the distance to its neighbor. Therefore, at the end of the initialization phase, each node in the network will also have estimates of the distances to its neighboring nodes. Later, when a sensor node collects data from the environment, it just chooses to send this data to the neighbor that is 1-hop closer to the sink than itself. If more than one neighbor is in this situation, the sensor node will send the data to the neighbor which is closer to itself, i.e., the neighbor from which it has the lowest distance estimation. We assume that during the data transmission phase the nodes can adjust their transmission power according to the estimated distance to their parent node in the routing tree, so that they transmit with "just enough" power to reach the parent node reliably. Depending on the transmission power that is used in the initialization phase, the resulting routing tree will be different. If a low transmission power is used, the nodes will be connected to the sink through several short hops; conversely, if the transmission power is higher, nodes will be connected through fewer, but longer, links.

This CTP-based tree construction protocol can be constructed and tuned in a very simple way, without the need of backwards transmissions from children to parents, and therefore maintains a very low initialization cost. The reason to propose this tree construction algorithm instead of using CTP trees directly is twofold. On the one hand, we are interested in using distance (not the ETX) for building the trees, as distance is likely to be more directly related to the data correlation. On the other hand, the parity of each node can be calculated very easily with the proposed algorithm, whereas with the CTP scheme it is more complex to determine, as a given path from a node to the sink can change several times during the initialization process, and this affects the parity of the node and its descendants. Note that node parity is important, since it is used to determine the role of each node in computing the wavelet transform.

3.2   Shortest Path Tree (SPT) and Minimum Distance Tree (MDT)

In order to broaden our evaluation of distributed data compression techniques over different routing trees, we also consider two other trees that can be constructed as well in a distributed way. The first one is a shortest path tree (SPT) that minimizes the sum of the squared distances between any node of the network and the sink. In an ideal propagation environment, the transmission cost for a given link is proportional to the square of the distance between the two nodes, so, under these circumstances, this shortest path tree is optimum in the sense that it minimizes the total transmission cost per bit. The second tree that we consider here is a variation of shortest path tree but, in contrast to the previous one, the objective function to be minimized in this case is the total distance to the sink. We name this tree minimum distance tree (MDT). Assuming that the correlation between sensor data is inversely proportional to the distance between the sensors, this tree is expected to perform well in terms of data compression. Both SPT and MDT can be computed in a distributed manner following the same approach used by CTP (substituting the cumulated ETX to the sink at each node, with the cumulated costs).

When the network configuration has to be built from scratch, both SPT/MDT and M-CTP can be seen as *greedy* procedures, as each node selects its parent *locally*, i.e. making a decision only on the basis of its neighbors' distances or accumulated costs. Specifically, in the initialization phase with SPT/MDT, each node n chooses a parent p_n which minimizes the sum of the costs from the parent to the sink, $C(p_n)$, plus the cost from node n to p_n, $C(n)$. When the costs are non-negative (as is the case of distances), minimizing *greedily* $C(p_n)+C(n)$ for each node n guarantees a globally optimal solution, as in fact this is nothing but the well-known Dijkstra algorithm. On the other hand, M-CTP simply minimizes the number of hops, by choosing closest matches when multiple equivalent choices are available, without any claim of optimality in the sense of total costs. The fundamental distinction between STP/MDT and M-CTP occurs when some tree *reconfiguration* is needed. Consider, without loss of generality, a simple example where a node n

with parity l in the tree runs out of battery and is no longer operational. Thus its "orphan" children need to be re-connected to other nodes. Note that in order to perform the wavelet transform this also implies that the parity of the children nodes needs to be updated to keep the transform consistent (the new configuration will be forwarded to the sink for the transform to be invertible, but we neglect this cost since it cannot be avoided with any of our routing schemes). Let $\rho(n)$ denote the set of children of node n, which have parity $l+1$ in the tree. When n dies, a reconfiguration procedure is launched. In the case of M-CTP, children nodes will have to look for a new parent node having depth in the tree greater than or equal to l using search procedure described in Section 3.1. If the new parent has level l, the configuration update is done; otherwise, if the level of the new parent is $l' > l$, each child $m \in \rho(n)$ set its parity to $l' + 1$ and the set of descendants $\rho(m)$, $\rho(\rho(m)) \ldots$, $m \in \rho(n)$ will modify their parity accordingly as well. Thus, in the worst case, a reconfiguration with M-CTP involves a propagation through descendants of parity level information, but *no other topological changes* in the set of descendants are carried out. In the case of SPT/MDT, on the contrary, each child node $m \in \rho(n)$ needs to look for a new candidate parent node p_m that minimizes $C(m) + C(p_m)$. But once the parent node is found, the descendants of m, $t \in \{\rho(m) \cup \rho(\rho(m)) \cup \ldots\}$, will not simply update their parity as in the M-CTP case, since each node t is supposed to be connected to the sink through the *minimum cost path* which minimizes $C(t) + C(p_t)$, where as before $C(p_t)$ denotes the cost of the path from the parent of t up to the sink. Therefore, a reconfiguration in the SPT/MDT case will in general entail a global re-computation of costs for the descendants of the modified node and in general this will lead to an overall modification of the tree. On the other hand, the proposed M-CTP algorithm can substantially reduce reconfiguration complexity, and thus it is a valid practical alternative to optimal SPT/MDT trees when network configuration changes frequently over time.

4   Results and Discussion

4.1   Description of Experimental Setup

In order to analyze precisely how the total energy consumption in the network depends on the different parameters we have run a set of simulations. In each simulation a network composed of 100 nodes (99 sensor nodes and one sink node) is deployed randomly in a $100\,m^2$ room. Three different data fields with different correlations (high, medium and low, denoted, H, M, L, respectively) are generated over the same region using a separable second order autoregressive filter. The communication channel is modeled using the lognormal shadowing path loss model [17], which establishes for each link a relation between the transmitted power (P_{TX}), the received power (P_{RX}) and the distance between transmitter and receiver (d):

$$P_{RX}(dBm) = P_{TX}(dBm) + A - 10\eta \log \frac{d}{d_0} + N. \tag{1}$$

A is a constant term that depends on the power loss for a reference distance d_0, η is the path loss exponent and $N \sim \mathcal{N}(0, \sigma^2)$ is a zero-mean gaussian noise

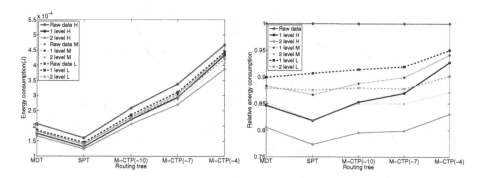

Fig. 1. Total energy consumption (*left*) and relative energy consumption with respect to the raw data case (*right*) for different routing trees. The labels M-CTP(x) correspond to the different M-CTP trees constructed with transmission power x. H, M and L represent the high, medium and low correlation datasets respectively.

with standard deviation σ. In the simulations we have set these parameters to $A = -80$ dB, $d_0 = 1$ m, $\eta = 2$ and $\sigma = 2$.

Different routing trees are constructed for each simulation: the SPT and MDT described in Section 3.2 and various M-CTP trees using the initialization procedure explained in Section 3.1 with different transmission powers. During this phase the nodes measure the RSS of all the packets they receive from their neighbors and they estimate the distances to them using the lognormal model. Then all the nodes send their data towards the sink and the tree-based wavelet transform is applied to reduce the amount of information that is sent over the network. During this data gathering phase the packets are sent with a transmission power that depends on the estimated inter-node distances: $P_{TX}(dBm) = S(dBm) - A + 10\eta \log(\frac{d}{d_0}) + M$, where S is the receiver sensitivity and M is a margin to reduce the packet loss probability (for the lognormal channel $M(dB) = -Q^{-1}(\text{PRP}) \cdot \sigma\sqrt{2}$, where Q^{-1} is the inverse Q function and PRP is the desired packet reception probability, in our simulations PRP=0.99). For the retransmissions it is reasonable to use a higher transmission power so, in the simulations, we have added 0.5 dB for each retransmission. After the data gathering phase, when the sink has received all the information, it computes the inverse transform so the original data is reconstructed (with some distortion).

4.2   Experimental Results

Using the described setup we have evaluated the total energy consumption in the network during the data gathering phase. Note that the cost of the initialization phase, i.e. the cost of constructing the routing tree, is not considered here. Fig. 1 shows the total energy consumption during the data collection phase for the tree routing algorithms described in Section 3. The cost of transmitting raw data across the network is compared against the cases of 1 and 2 levels of wavelet decomposition. For the tunable M-CTP trees, it can be seen that the

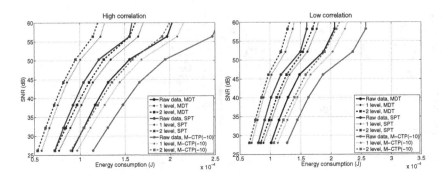

Fig. 2. SNR–Energy consumption for two datasets with different correlation. Each point in the curves represents the average over 50 simulations.

energy consumption generally increases with the transmission power; at the same time, the relative cost reduction with respect to simple data gathering (raw bits transmission) tends to get smaller. For the MDT and the SPT the results are better than those obtained with the M-CTP trees, as the former trees have a higher number of (shorter) hops. This supports the thesis that longer paths with shorter hops are better from both a transform coding gain point of view and for routing efficiency. Note that the consumption reduction yielded by wavelet transform with respect to raw data gathering does not change across the different M-CTP trees as significatively as does the total cost. This is partially due to an intrinsic overhead of the adopted transform, which necessarily requires the transmission of non compressed information for 1 or 2 hops. However, we believe that this reduced gain could improve in the case of a denser network[1].

An overall evaluation of in-network compression also needs to consider the quality of the data reconstructed at the sink. We consider in Fig. 2 SNR–Energy consumption curves for the three different routing trees (only the M-CTP tree with a transmission power of -10 dBm is shown in the graph). The SNR here is the reconstruction signal-to-quantization noise. Clearly, to obtain a lower distortion in the reconstructed data, the energy to be consumed is larger (as the number of bits increase). However, we observe that, e.g. for a target distortion of 40 dB, the M-CTP tree yields a cost reduction of about 10% (14%) with a 1 level wavelet transform in the case of data with low (high) correlation; using 2 levels of decomposition we can further gain 5% (6%) for the case of low (high) correlation. These gains are basically the same for the MDT tree, while the SPT produces a further transform gain of about 1-2%. Note that in terms of absolute consumption-distortion performance, the SPT gives the best performance, as it is designed to provide the cost-optimal tree.

---

[1] We are not considering channel capacity and interference effects, for which having a denser network implies an asymptotically increasing cost due to retransmission and reduced throughput [18].

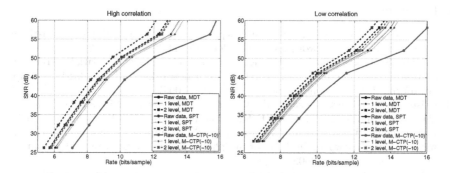

Fig. 3. SNR–Rate curves for two datasets with different correlation. Each point in the curves represents the average over 50 simulations.

In order to separate the specific contribution of the transform to the total cost, we plot in Fig. 3 SNR–Bit rate curves for the same routing trees of Fig. 2 (we only show the performance of one of the M-CTP routing trees for the sake of clarity). Again, the best performance is obtained with the 2-level transform. The marginal gain of using a transform in the MDT with respect to M-CTP is on average about 0.5 dB for both the high and low correlation datasets. The SPT can benefit more from higher correlation, and the gain when the transform is applied increases to 3 dB for the high correlation data (while it stops at 1.5 dB for low correlation). This suggests that a tree which is optimal for routing is in general optimal also for transform coding, even if a sub-optimal tree like the M-CTP can be still preferable in practice when other non-functional requirements (such as the reconfiguration time) come into play.

5 Conclusions

This paper aims at corroborating the *practical* feasibility of transform-based in-network compression using a *totally distributed* initialization approach. Our main contribution is to propose a simple, yet effective distributed tree-construction algorithm based on CTP, which fits particularly well to the 2D wavelet transform adopted in this work by embedding, explicitly, the distance between nodes and thus data correlation. We discuss how to tune, in a practical situation, the initialization power of the M-CTP scheme in order to produce trees with different depth, and we show that deeper trees will lead in general to better performance in terms of transmission costs. The proposed scheme entails lower initialization costs in comparison to distributed versions of optimal routing trees such as the SPT when some nodes need to be reconfigured (which is particularly beneficial in case of highly dynamic networks).

References

1. Scaglione, A., Servetto, S.: On the interdependence of routing and data compression in multi-hop sensor networks. Wireless Networks 11(1), 149–160 (2005)
2. Pattem, S., Krishnamachari, B., Govindan, R.: The impact of spatial correlation on routing with compression in wireless sensor networks. In: Proc. 3rd Int. Symp. on Information Processing in Sensor Networks (2004)
3. Cristescu, R., Beferull-Lozano, B., Vetterli, M.: Networked Slepian-Wolf: theory, algorithms, and scaling laws. IEEE Trans. on Inf. Theory 51(12), 4057–4073 (2005)
4. Krishnamachari, B., Estrin, D., Wicker, S.B.: The impact of data aggregation in wireless sensor networks. In: Proc. of the 22nd Int. Conf. on Distributed Computing Systems, pp. 575–578 (2002)
5. Servetto, S.: Distributed signal processing algorithms for the sensor broadcast problem. In: Proc. 37th Annual Conf. Inform. Sciences Syst. (March 2003)
6. Ciancio, A., Pattem, S., Ortega, A., Krishnamachari, B.: Energy-efficient data representation and routing for wireless sensor networks based on a distributed wavelet compression algorithm. In: Proc. 5th Int. Conf. on Information Processing in Sensor Networks, April 2006, pp. 309–316 (2006)
7. Shen, G., Ortega, A.: Optimized distributed 2D transforms for irregularly sampled sensor network grids using wavelet lifting. In: Proc. IEEE Int. Conf. on Acoustics, Speech and Signal Processing, April 2008, pp. 2513–2516 (2008)
8. Pattem, S., Shen, G., Chen, Y., Krishnamachari, B., Ortega, A.: Senzip: an architecture for distributed en-route compression in wireless sensor networks. In: ACM/IEEE Int. Conf. on Information Processing in Sensor Networks (April 2009)
9. TinyOS-2: Collection tree protocol, http://www.tinyos.net/tinyos-2.x/doc/
10. Cristescu, R., Beferull-Lozano, B., Vetterli, M.: On network correlated data gathering. In: Proc. 23rd Annual Joint Conf. of the IEEE Computer and Communications Societies, March 2004, vol. 4 (2004)
11. Ciancio, A., Ortega, A.: A dynamic programming approach to distortion-energy optimization for distributed wavelet compression with applications to data gathering inwireless sensor networks. In: Proc. IEEE Int. Conf. on Acoustics, Speech and Signal Processing (April 2006)
12. Wagner, R., Choi, H., Baraniuk, R., Delouille, V.: Distributed wavelet transform for irregular sensor network grids. In: Proc. IEEE Stat. Sig. Proc. Workshop (July 2005)
13. Duarte, M., Wakin, M., Baron, D., Baraniuk, R.: Universal distributed sensing via random projections. In: Proc. 5th Int. Conf. on Information processing in sensor networks, April 2006, pp. 177–185 (2006)
14. Acimovic, J., Beferull-Lozano, B., Cristescu, R.: Adaptive distributed algorithms for power-efficient data gathering in sensor networks. In: Proc. Int. Conf. on Wireless Networks, Communications and Mobile Computing (June 2005)
15. Shen, G., Ortega, A.: Joint routing and 2D transform optimization for irregular sensor network grids using wavelet lifting. In: Proc. 7th Int. Conf. on Information Processing in Sensor Networks (April 2008)
16. Shen, G., Pattem, S., Ortega, A.: Energy-efficient Graph-based Wavelets for Distributed Coding in Wireless Sensor Networks. In: Proc. IEEE Int. Conf. on Acoustics, Speech and Signal Processing (April 2009)
17. Rappaport, T.: Wireless communications. Prentice Hall PTR, Englewood Cliffs (2002)
18. Marco, D., Duarte-Melo, E., Liu, M., Neuhoff, D.: On the many-to-one transport capacity of a dense wireless sensor network and the compressibility of its data. In: Zhao, F., Guibas, L.J. (eds.) IPSN 2003. LNCS, vol. 2634, pp. 1–16. Springer, Heidelberg (2003)

Spatially-Localized Compressed Sensing and Routing in Multi-hop Sensor Networks[*]

Sungwon Lee, Sundeep Pattem, Maheswaran Sathiamoorthy,
Bhaskar Krishnamachari, and Antonio Ortega

Dept. of Electrical Engineering, University of Southern California,
Los Angeles, CA 90089, USA
{sungwonl,pattem,msathiam,bkrishna,antonio.ortega}@usc.edu

Abstract. We propose energy-efficient compressed sensing for wireless sensor networks using spatially-localized sparse projections. To keep the transmission cost for each measurement low, we obtain measurements from clusters of adjacent sensors. With localized projection, we show that joint reconstruction provides significantly better reconstruction than independent reconstruction. We also propose a metric of energy overlap between clusters and basis functions that allows us to characterize the gains of joint reconstruction for different basis functions. Compared with state of the art compressed sensing techniques for sensor network, our simulation results demonstrate significant gains in reconstruction accuracy and transmission cost.

1 Introduction

Joint routing and compression has been studied for efficient data gathering of locally correlated sensor network data. Most of the early works were theoretical in nature and, while providing important insights, ignored the practical details of how compression is to be achieved [1,2,3]. More recently, it has been shown how practical compression schemes such as distributed wavelets can be adapted to work efficiently with various routing strategies [4,5,6].

Existing transform-based techniques, including wavelet based approaches [4,5,7] and the distributed KLT [8], can reduce the number of bits to be transmitted to the sink thus achieving overall power savings. These transform techniques are essentially critically sampled approaches, so that their cost of gathering scales up with the number of sensors, which could be undesirable when large deployments are considered. Compressed sensing (CS) has been considered as a potential alternative in this context, as the number of samples required (i.e., number of sensors that need to transmit data), depends on the characteristics (sparseness) of the signal [9,10,11].

---

[*] The work described here is supported in part by NSF through grants CNS-0347621, CNS-0627028, CCF-0430061, CNS-0325875, and by NASA through AIST-05-0081. Any opinions, findings, and conclusions or recommendations expressed in this material are those of the author(s) and do not necessarily reflect the views of the NSF or NASA.

N. Trigoni, A. Markham, and S. Nawaz (Eds.): GSN 2009, LNCS 5659, pp. 11–20, 2009.
© Springer-Verlag Berlin Heidelberg 2009

In addition CS is also potentially attractive for wireless sensor networks because most computations take place at the decoder (sink), rather than encoder (sensors), and thus sensors with minimal computational power can efficiently encode data.

However, while the potential benefits of CS have been recognized [12,13], significant obstacles remain for it to become competitive with more established (e.g., transform-based) data gathering and compression techniques. A primary reason is that CS theoretical developments have focused on *minimizing the number of measurements* (i.e., the number of samples captured), rather than on *minimizing the cost of each measurement*. In many CS applications (e.g., [14] [15]), each measurement is a linear combination of many (or all) samples of the signal. It is easy to see why this is not desirable in the context of a sensor network: the signal to be sampled is *spatially distributed* so that measuring a linear combination of all the samples would entail a significant transport cost to generate each aggregate measurement. To address this problem, *sparse measurement* approaches (where each measurement requires information from a few sensors) have been proposed for both single hop [16] and multi-hop [12,13] sensor networks.

In this paper our goal is to make explicit the trade-off between measurement cost and reconstruction quality. We note that lowering transport costs requires spatially localized gathering. Signals to be measured can be expressed in terms of elementary basis functions. We show that the performance of CS greatly depends on the nature of these bases (in particular, whether or not they are spatially localized). Thus, the specific data gathering strategy will depend in general on the signals to be measured. We propose a novel spatially-localized projection technique based on clustering groups of neighboring sensors. The idea of the local projection is similar to the localized mode in [17] which exploits correlations among multiple nodes in clusters rather than the sparseness of data used in our approach. We show that joint reconstruction of data across clusters leads to significant gains over independent reconstruction. Moreover, we show that reconstruction performance depends on the level of "overlap" between these data-gathering clusters and the elementary basis on which the signals are represented. We propose methods to quantify this spatial overlap, which allow us to design efficient clusters once the bases for the signal are known. Our simulation results demonstrate significant gains over state of the art CS techniques for sensor networks [16,13].

The remainder of this paper is organized as follows. Section 2 presents CS basics and motivation. Section 3 introduces spatially-localized CS. Section 4 presents efficient clustering for spatially-localized CS and Section 5 provides simulation results, comparing proposed method and previously proposed CS techniques [16,13]. Section 6 concludes the paper.

2   Background and Motivation

Compressed Sensing (CS) builds on the observation that an n-sample signal (x) having a sparse representation in one basis can be recovered from a small number of projections (smaller than n) onto a second basis that is incoherent with the first [9,10]. If a signal, $x \in \Re^n$, is sparse in a given basis Ψ

(the sparsity inducing basis), $x = \Psi a, |a|_0 = k$, where $k \ll n$, then we can reconstruct the original signal with $O(k log n)$ measurements by finding sparse solutions to under-determined, or ill-conditioned, linear systems of equations, $y = \Phi x = \Phi \Psi x = H x$, where H is known as the holographic basis. Reconstruction is possible by solving the convex unconstrained optimization problem, $\min_x \frac{1}{2} \|y - Hx\|_2^2 + \gamma \|x\|_1$, if Φ and Ψ are mutually incoherent [11]. The mutual coherence, $\mu(\Phi \Psi) = \max_{k,j} |\langle \phi_k, \psi_j \rangle|$, serves as a rough characterization of the degree of similarity between the sparsity and measurement systems. For μ to be close to its minimum value, each of the measurement vectors must be spread out in the Ψ domain.

Measurements, y_i, are projections of the data onto the measurement vectors, $y_i = \langle \phi_i, x \rangle$, where ϕ_i is the ith row of Φ. Interestingly, independent and identically distributed (i.i.d.) Gaussian, Rademacher (random ± 1) or partial Fourier vectors provide useful universal measurement bases that are incoherent with any given Ψ with high probability. The measurement systems covered in traditional compressed sensing are typically based on these kinds of "dense" matrices, i.e., there are very few zero entries in Φ.

The dense projections of traditional compressed sensing are not suitable for sensor networks due to their high energy consumption. With a dense projection, every sensor is required to transmit its data once for each measurement, so the total cost can potentially be higher than that of a raw data gathering scheme. If the number of samples contributing to each measurement decreases, the cost is reduced by a factor that depends on the sparsity of the measurement matrix (see [12,13] for an asymptotic analysis for different measurement matrices.) Note, however, that in addition to sparsity, the gathering cost also depends on the position of the sensors whose samples are aggregated in the measurements. If sensors contributing to a given measurement are far apart, the cost will still be significant even with a sparse measurement approach. This is our main motivation to develop spatially-localized sparse projections.

3   Spatially-Localized Compressed Sensing

3.1   Low-Cost Sparse Projection Based on Clustering

In order to design distributed measurements strategies that are both sparse *and* spatially localized, we propose dividing the network into clusters of adjacent nodes and forcing projections to be obtained only from nodes within a cluster. As an example, in this paper we consider two simple clustering approaches. For simplicity, we assume that all clusters contain the same number of nodes. When N_c clusters are used, each cluster will contain $\frac{N}{N_c}$ nodes. In "square clustering", the network is partitioned into a certain number of equal-size square regions. Alternatively, in "SPT-based clustering", we first construct shortest path tree (SPT) then, based on that, we iteratively construct clusters from leaf nodes to the sink. If incomplete clusters encounter nodes where multiple paths merge, we group them into a complete cluster under the assumption that nodes sharing a common parent node are likely to be close to each other.

Any chosen clustering scheme can be represented in CS terms by generating the corresponding measurement matrix, $\boldsymbol{\Phi}$, and using it to reconstruct the original signal. Each row of $\boldsymbol{\Phi}$ represents the aggregation corresponding to one measurement: we place non-zero (or random) coefficients in the positions corresponding to sensors that provide their data for a specific measurement and the other positions are set to zero. Thus, the sparsity of a particular measurement in $\boldsymbol{\Phi}$ depends on the number of nodes participating in this aggregation.

For simplicity, we consider non-overlapped clusters with the same size. This leads to a block-diagonal structure for $\boldsymbol{\Phi}$. Note that recent work [18] [19], seeking to achieve fast CS computation, has also proposed measurement matrices with a block-diagonal structure, with results comparable to those of dense random projections. Our work, however, is motivated by achieving spatially localized projections so that our choice of block-diagonal structure will be constrained by the relative positions of the sensors (each block corresponds to a cluster).

3.2   Sparsity-Inducing Basis and Cluster Selection

While it is clear that localized gathering leads to lower costs, it is not obvious how it may impact reconstruction quality. Thus, an important goal of this paper is to study the interaction between localized gathering and reconstruction. A key observation is that in order to achieve both efficient routing and adequate reconstruction accuracy, the structure of the sparsity-inducing basis should be considered. To see this, consider the case where signals captured by the sensor network can be represented by a "global" basis, e.g., DCT, where each basis spans all the sensors in the network. Then the optimally incoherent measurement matrix will be the identity matrix, $\boldsymbol{I}$, thus a good measurement strategy is simply to sample $k \log n$ randomly chosen sensors and then forward each measurement directly to the sink (no aggregation). Alternatively, for a completely localized basis, e.g., $\boldsymbol{\Psi} = \boldsymbol{I}$, a dense projection may be best. However, once the transport costs have been taken into account, it is better to just have sensors transmit data to the sink via the SPT whenever they sense something "new" (e.g., when measurements exceed a threshold). In other words, even if CS theory suggests a given type of measurements (e.g., dense projection for the $\boldsymbol{\Psi} = \boldsymbol{I}$ case), applying these directly may not lead to an efficient routing and therefore efficient distributed CS may not be achievable.

In this paper we consider intermediate cases, in particular those where localized bases with different spatial resolutions are considered (e.g., wavelets). Candes *et al.* [11] have shown that a partial Fourier measurement matrix is incoherent with wavelet bases at fine scales. However, such a dense projection is not suitable for low-cost data gathering for the reasons discussed above. Next we explore appropriate spatially-localized gathering for data that can be represented in localized bases such as wavelets.

4   Efficient Clustering for Spatially-Localized CS

4.1   Independent vs. Joint Reconstruction

To study what clustering scheme is appropriate for CS, we first compare two types of reconstruction: independent reconstruction and joint reconstruction. Suppose that we construct a set of clusters of nodes and collect a certain number of local measurements from each cluster. With a given clustering/localized projection, joint reconstruction is performed with the basis where sparseness of signal is originally defined while independent reconstruction is performed with truncated basis functions corresponding to each cluster.

Equation (1) describes an example for two clusters with the same size. ψ_1 and ψ_1 correspond to localized projections in each cluster. For joint reconstruction, the original sparsity inducing basis, Ψ, is employed. But, for independent reconstruction, data in the first cluster are reconstructed with partial basis functions, ψ_1 and ψ_2, and those in the second cluster are with ψ_3 and ψ_4 thus, when N_c clusters are involved, independent reconstruction should be performed N_c times, once for each cluster.

$$H = \Phi\Psi = \begin{bmatrix} \phi_1 & 0 \\ 0 & \phi_2 \end{bmatrix} \begin{bmatrix} \psi_1 & \psi_2 \\ \psi_3 & \psi_4 \end{bmatrix} \Rightarrow \begin{cases} H_1 = [\phi_1\psi_1, \phi_1\psi_2] \\ H_2 = [\phi_2\psi_3, \phi_2\psi_4] \end{cases} \qquad (1)$$

Joint reconstruction is expected to outperform independent reconstruction because the sparseness of data is determined by the original basis, Ψ. Also, with joint reconstruction, measurements taken from a cluster can also convey information about data in other clusters because basis functions overlapped with more than one clusters can be identified with measurements from those clusters.

As basis functions are overlapped with more clusters, joint reconstruction has potentially higher chance to reconstruct signal correctly. To augment gains from localized CS with clustering scheme, how to choose the clustering should be based on the basis, so that overlap is encouraged. The degree of overlapping between basis functions and clusters can be measured in many different ways. One possible approach is to measure energy of basis functions captured by each cluster.

4.2   Energy Overlap Analysis

To characterize the distribution of energy of basis functions with respect to the clusters, we present a metric, E_{oa}, and an analysis of the worst-case scenario. We assume that signal is sparse in $\Psi \in \Re^{n \times n}$. And, $\psi(i,j)$ corresponds to the jth entry in the ith column of Ψ and each column is normalized to one. Suppose that N_c is the number of clusters and C_i is a set of nodes contained in the ith cluster. The energy overlap between the ith cluster and the jth basis vector, $E_o(i,j)$, is

$$E_o(i,j) = \sum_{k \in C_i} \psi(j,k)^2 \qquad (2)$$

Energy overlap per overlapped basis, E_{oa}. Average energy overlap per overlapped basis is a good indicator of distribution of energy of basis functions. For each cluster, $E_{oa}(i)$ is computed as

$$E_{oa}(i) = \frac{1}{N_o(i)} \sum_{j=1}^{N} E_o(i,j), \ \forall i \in \{1, 2, \cdots, N_c\}, \tag{3}$$

where $N_o(i)$ is the number of basis functions overlapped with the ith cluster. Then, we compute E_{oa} by taking average of $E_{oa}(i)$ over all clusters, $E_{oa} = \frac{1}{N_c} \sum_{i=1}^{N_c} E_{oa}(i)$. Intuitively, this metric shows how much energy of basis functions are captured by each cluster. Thus, as energy of basis functions are more evenly distributed over overlapped clusters, E_{oa} decreases, which leads to better reconstruction performance with joint reconstruction. If the specific basis contributing a lot to the cluster is not in the data support, the measurements from this cluster does not notably increase the reconstruction performance.

Worst case analysis. It would be useful to have a metric to determine the number of projections required from each local cluster in order to achieve a certain level of reconstruction performance. We first define what the worst-case is, then try to characterize the 'worst-case scenario' performance.

With the global sparsity of K, the worst case scenario is when all K basis vectors supporting data are completely contained in a single cluster. Since the identity of this cluster is not known *a priori* and projections from other clusters not overlapped with those basis vectors do not contribute to reconstruction performance as much as projections from that cluster, $O(K)$ projections would be required from each cluster. But, note that, in general, the coarsest basis vector representing DC component is likely to be overlapped with more than one cluster and chosen as data support for real signal. Thus, in practice, performance could be better than our estimation.

To analyze the worst-case scenario, we assume that we know the basis functions, clustering scheme and the value of K *a priori*. For each cluster, we first choose K basis functions with highest energy overlap with the cluster. Then, we compute the sum of the energy overlap of the chosen basis functions. To simplify analysis, we take the average over all clusters. Minimum number of measurements for each cluster indirectly depends on overlap energy in the worst-case scenario. For example, with DCT basis and four clusters with the same size, overlap energy for each cluster is equal to $\frac{K}{4}$ in the worst-case thus the total number of measurements will be $O(K)$.

5   Simulation Results

For our simulation, we used 500 data generated with 55 random coefficients in different basis. In the network, 1024 nodes are deployed on the square grid and error free communication is assumed. Two types of clustering are considered: square-clustering and SPT-based clustering. We do not assume any priority is given to specific clusters for measurements, i.e., we collect the same number of localized measurements for each cluster. With localized projection in each cluster, data is reconstructed jointly or independently with Gradient Pursuit for Sparse Reconstruction (GPSR) [20] provided in [21]. To evaluate performance, SNR is used to evaluate reconstruction accuracy. For cost evaluation, transmission cost is computed by $\sum (bit) \times (distance)^2$, as was done in [5], but the work could be extended to use more realistic cost metrics. The cost ratio in our simulation is the ratio to the cost for raw data gathering without compression.

To compare independent reconstruction with joint reconstruction, we used square-clustering scheme with two different number of clusters and Haar basis with decomposition level of 5. In Fig. 1(a), DRP corresponds to the case that takes 256 global measurements from all the nodes in the network then reconstructs data with joint reconstruction. Other curves are generated from localized measurements in each cluster and the two types of reconstruction are applied respectively.

Fig. 1 (a) shows that joint reconstruction outperforms independent reconstruction. As discussed in Section 4.1, joint reconstruction can alleviate the worst situation by taking measurements from other clusters overlapped with basis functions in the data support. In following simulation, all the data was jointly reconstructed.

With joint reconstruction and Haar basis, Fig. 1 (b) shows that SPT-based clustering outperforms square clustering for different number of clusters (N_c). As N_c increases, reconstruction accuracy decreases because measurement matrix becomes sparser as network is separated into more equal-size clusters. However, once the transport costs have been taken into account, more clusters show better performance because cost per each measurement decreases. Since we also observed this trend for different bases, we will focus on 64 SPT-based clusters in following simulation.

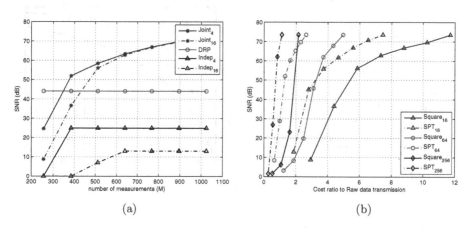

(a)   (b)

Fig. 1. Comparison of different types of reconstruction and clustering scheme with Haar basis with decomposition level of 5. (a) Comparison of independent reconstruction and joint reconstruction. (b) Cost ratio to raw data gathering vs. SNR with different number of clusters and clustering schemes.

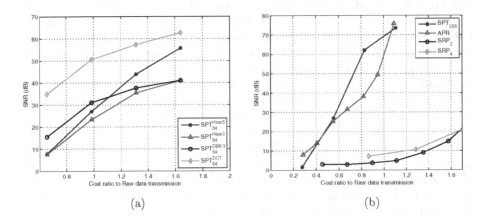

(a)   (b)

Fig. 2. Performance comparison in terms of cost ratio to raw data gathering vs. SNR (a) for different basis functions and 64 SPT-based clusters. (b) for 256 SPT-based clusters with other CS approaches with Haar basis with level of decomposition of 5.

To investigate effects of different bases, we consider the joint reconstruction performance with different basis functions with 64 SPT-based clusters: 1) DCT basis, where each basis vectors have high overlaps in energy which distributed throughout the network 2) Haar basis, where the basis vectors have less overlap and the energy distribution varies from being very localized to global for different basis vectors and 3) Daubechies (DB6) basis, where the overlaps and distribution are intermediate to DCT and Haar. The result in Fig. 2 (a) confirms our intuition.

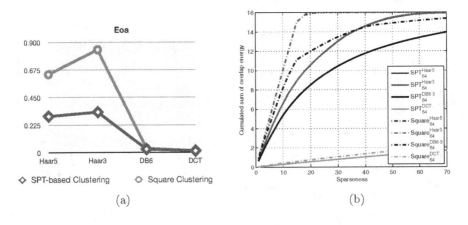

(a)   (b)

Fig. 3. E_{oa} and worst-case analysis with different basis and clustering schemes. Smaller values of metrics indicate more even distribution of overlap energy thus better reconstruction. (a) E_{oa}; values are average over clusters and variances are ignored because they are relatively small. (b) Worst-case analysis; average of cumulated overlap energy with increasing number of basis.

Thus, for the same clustering scheme, the gains from joint reconstruction depend on how "well-spread" the energy in the basis vectors is.

As an indicator of distribution of energy of basis functions, E_{oa} is computed with different clustering schemes for different basis functions. Fig 3 (a) shows that E_{oa} accurately distinguishes performance between two different clusters; SPT-based clusters capture more energy of basis functions than square clusters then lead to better reconstruction. For different basis functions, E_{ov} shows lower overlap energy as basis functions are more spread over in spatial domain.

The result for worst-case analysis is shown in Fig. 3 (b). The results show that, for the same basis function, SPT-based clustering reduces more energy than square clustering thus requires fewer measurements for each cluster. For DCT basis, the energy increases slower than other basis because energy of DCT basis is evenly spread over all clusters. For Haar basis, overlap energy increases very sharply with a few basis functions then it is saturated because energy of some basis functions is concentrated in small regions. With Daubechies basis, the energy is somewhat between two previous bases as we expected.

Fig. 2(b) shows that our approach outperforms other CS approaches [16,13]. APR corresponds to a scheme that aggregation occurs along the shortest path to the sink and all the sensors on the paths provide their data for measurements. SRP with different parameter, $s' = s \times n$, represents a scheme that randomly chooses s' nodes without considering routing then transmit data to the sink via SPT with opportunistic aggregation.

In the comparison, SRP performs worse than the others because, as we expected, taking samples from random nodes for each measurement significantly increases total transmission cost. Our approach and APR are comparable in terms of transmission cost but our approach shows better reconstruction. The performance gap is well explained by N_{oa} for APR and our approach: 0.247 and 0.171 respectively. Lower N_{oa} indicates that energy of basis functions are more evenly distributed over overlapped clusters thus those functions are more likely to be identified with joint reconstruction.

6 Conclusion

We have proposed a framework for efficient data gathering in wireless sensor network by using spatially-localized compressed sensing. With localized projection in each cluster, joint reconstruction has shown better performance than independent reconstruction because joint reconstruction can exploit measurements in multiple clusters, corresponding to energy in a given basis function that overlaps those clusters. Our proposed approach outperforms over state of the art CS techniques [16,13] because our method achieves power savings with localized aggregation and captures more evenly distributed energy of basis functions. Moreover, we proposed methods to quantify the level of "energy overlap" between the data gathering clusters and the elementary basis on which the signals are represented, which allows us to to design efficient clusters once the bases for the signal are known. Based on the metric, we hope to design an optimal clustering scheme in near future.

References

1. Cristescu, R., Beferull-Lozano, B., Vetterli, M.: On network correlated data gathering. In: INFOCOM (March 2004)
2. Pattem, S., Krishnamachari, B., Govindan, R.: The impact of spatial correlation on routing with compression in wireless sensor networks. In: IPSN (April 2004)
3. von Rickenbach, P., Wattenhofer, R.: Gathering correlated data in sensor networks. In: DIALM-POMC. ACM, New York (2004)
4. Ciancio, A., Pattem, S., Ortega, A., Krishnamachari, B.: Energy-efficient data representation and routing for wireless sensor networks based on a distributed wavelet compression algorithm. In: IPSN (April 2006)
5. Shen, G., Ortega, A.: Joint routing and 2d transform optimization for irregular sensor network grids using wavelet lifting. In: IPSN (April 2008)
6. Pattem, S., Shen, G., Chen, Y., Krishnamachari, B., Ortega, A.: Senzip: An architecture for distributed en-route compression in wireless sensor networks. In: ESSA (April 2009)
7. Wagner, R., Choi, H., Baraniuk, R., Delouille, V.: Distributed wavelet transform for irregular sensor network grids. In: SSP (July 2005)
8. Gastpar, M., Dragotti, P., Vetterli, M.: The distributed karhunen-loeve transform. In: MMSP (December 2002)
9. Donoho, D.L.: Compressed sensing. IEEE Transactions on Information Theory (April 2006)
10. Candes, E., Romberg, J., Tao, T.: Robust uncertainity principles: exact signal reconstruction from highly incomplete frequency information. IEEE Transactions on Information Theory (February 2006)
11. Candes, E., Romberg, J.: Sparsity and incoherence in compressive sampling. Inverse Problems (June 2007)
12. Lee, S., Pattem, S., Sathiamoorthy, M., Krishnamachari, B., Ortega, A.: Compressed sensing and routing in sensor networks. USC CENG Technical Report (April 2009)
13. Quer, G., Masierto, R., Munaretto, D., Rossi, M., Widmer, J., Zorzi, M.: On the interplay between routing and signal representation for compressive sensing in wireless sensor network. In: ITA (February 2009)
14. Lustig, M., Donoho, D., Pauly, J.M.: Sparse MRI: The application of compressed sensing for rapid MR imaging. In: MRM (December 2007)
15. Duarte, M.F., Davenport, M.A., Takhar, D., Laska, J.N., Sun, T., Kelly, K.F., Baraniuk, R.G.: Single pixel imaging via compressive sampling. IEEE Signal Processing Magazine (March 2008)
16. Wang, W., Garofalakis, M., Ramchandran, K.: Distributed sparse random projections for refinable approximation. In: IPSN (April 2007)
17. Deligiannakis, A., Kotidis, Y., Roussopoulos, N.: Dissemination of compressed historical information in sensor networks. VLDB Journal (2007)
18. Gan, L., Do, T.T., Tran, T.D.: Fast compressive imaging using scrambled block hadamard ensemble (preprint, 2008)
19. Do, T., Tran, T., Gan, L.: Fast compressive sampling with structurally random matrices. In: ICASSP (April 2008)
20. Figueiredo, M., Nowak, R., Wright, S.: Gradient projection for sparse reconstruction: application to compressed sensing and other inverse problems. IEEE Journal of Selected Topics in Signal Processing (2007)
21. Figueiredo, M., Nowak, R., Wright, S.: Gradient projection for sparse reconstruction (January 2009), http://www.lx.it.pt/~mtf/GPSR/

Estimation of Pollutant-Emitting Point-Sources Using Resource-Constrained Sensor Networks

Michael Zoumboulakis and George Roussos

School of Computer Science and Information Systems Birkbeck College,
University of London, Malet Street, London WC1E 7HX, UK
{mz,gr}@dcs.bbk.ac.uk

Abstract. We present an algorithm that makes an appropriate use of a Kalman filter combined with a geometric computation with respect to the localisation of a pollutant-emitting point source. Assuming resource-constrained inexpensive nodes and no specific placement distance to the source, our approach has been shown to perform well in estimating the coordinates and intensity of a source. Using local gossip to directionally propagate estimates, our algorithm initiates a real-time exchange of information that has as an ultimate goal to lead a packet from a node that initially sensed the event to a destination that is as close to the source as possible. The coordinates and intensity measurement of the destination comprise the final estimate. In this paper, we assert that this low-overhead coarse localisation method can rival more sophisticated and computationally-hungry solutions to the source estimation problem.

Keywords: Source Localisation, Spatial Event Detection, Estimation, Kalman Filters.

1 Introduction

We live in days of elevated risk that terrorists could acquire Chemical, Biological and Radiological (CBR) weapons to attack major cities. In a world of easy availability of CBR raw materials in hospitals of failed states, the greatest concern for citizens of urban centres is not over an attack by a nuclear warhead but with a "dirty-bomb" that would contaminate a wide area, trigger widespread panic and cause severe disruption [12]. Government agencies, such as the Home Office in the UK, are involved in "resilience" programmes, set to cost GBP 3.5bn per year by 2011, with the aim to *"ensure that, in the event of a terrorist incident the response from all concerned will be quick and effective, so that lives can be saved and the impact on property and the environment minimised"* [15].

The principal concern of such programmes is the detailed response strategy in the case of a CBR attack. For this, accurate assessment and modelling of hazards [5] are essential. Modelling can be split in the following two sub-problems: forward and inverse. While the forward problem involves the dispersion prediction and hazard assessment given a known location of a point source emitting pollutants in space, the inverse problem involves locating this source given some

N. Trigoni, A. Markham, and S. Nawaz (Eds.): GSN 2009, LNCS 5659, pp. 21–30, 2009.
© Springer-Verlag Berlin Heidelberg 2009

measurements of the pollutants in the atmosphere. The quick localisation of an attack source can assist the specialist personnel in neutralising the threat as well as predicting the dispersion cloud and evacuating citizens away from it.

In this paper, we consider a scenario where a number of low-cost, resource-constrained sensor nodes, such as the TMote Sky [3], are deployed in an urban area with the task of detecting the presence of certain pollutants in the atmosphere. There exist real-world deployments, such as the U.S. Federal Sensor-Net's [16] testbeds at Washington DC and New York City, that aim to serve the aforementioned objective. Following the detection of the pollutant — a problem addressed by our earlier work on event detection [23] — the goal is to compute an estimate of the source's location and intensity.

We use an iterative in-network approach that does not rely on powerful nodes or network-wide collection and offline processing. Our method involves a pre-determined number of *initiating* nodes that independently sense the spatial event. Each of these nodes, begins a procedure that aims to route a packet intelligently towards the source of the spatial event. The ultimate aim is that after a small number of hops the packet should be at a node very close to the source of the spatial event — we will call this the *destination* node. By taking the location coordinates and measurement at the destination node, we have a coarse estimate of the source's location and intensity.

The method is designed to operate as close to real-time as possible and to be lightweight in terms of network communication. To address the latter point, we use directional local gossip to propagate event state information. This is somewhat akin to the Trickle [10] algorithm that uses polite gossip to bring network consistency with respect to a set of global shared parameters. Our algorithm starts with the initiating node making a guess of the spatial event state at single-hop neighbouring nodes and sending it to the local broadcast address. Nodes hearing the broadcast reply with their own measurements. The estimation error is calculated and the node that minimises the error is selected as the next hop and its measurement is used to correct the initial prediction made. This "predict-correct" cycle is in fact a straightforward Kalman filter. The node selected as next hop receives the filter parameters and repeats the steps. The procedure is continued iteratively, recording the network path along the way and it exits when a node runs out of unvisited neighbours that are likely to lead any closer to the source. The exact details of the algorithm will be described in section 2.2.

In the remainder of this paper, we will formalise the problem and we will discuss the requirements for a solution. We will then present our algorithm for decentralised in-network point-source estimation together with discussion of a test case and experimental simulation results. We will conclude by reviewing selected related work together with future plans.

1.1   Problem Statement and Requirements

We consider a single static point source that emits a pollutant of chemical or radiological nature. This source is located at the unknown coordinates (x_s, y_s) in the two dimensional coordinate system. The presence and intensity of the pollutant in the atmosphere is sensed by N sensor nodes located at (x_i, y_i) coordinates with $i = 1, 2, \ldots, N$. The goal is to devise a computational method that estimates the coordinates $(\hat{x}_s, \hat{y}_s)$ and intensity of the source, within some error ϵ.

Formally, given z_i sensor measurements collected at coordinates (x_i, y_i) the goal is to estimate the vector $\hat{v}$:

$$\begin{bmatrix} \hat{x}_s & \hat{y}_s & \hat{I} \end{bmatrix}^T$$

where $\hat{I}$ is the intensity estimate as it would be measured 1 meter from the source [5].

The complexity of a proposed solution for this problem varies significantly depending on the assumptions made. In our case we assume a single static point source in $\mathbb{R}^2$ and a steady-state dispersion model. The latter point refers to a simplified gas concentration model, adapted from [7], where measurements are time-averaged and constant with respect to time. Furthermore, we assume that the nodes are aware of their own location coordinates.

The solution to the source localisation problem needs to be lightweight both in terms of computation and communication. It needs to operate in a decentralised manner without relying upon powerful nodes or offline processing and it needs to be capable of converging to an estimate rapidly — typically under two minutes.

Our approach tolerates both non-uniform distributions and, due to the fact that our estimation algorithm employs a Kalman Filter, a degree of measurement and process noise.

2   Iterative Source Location Estimation

2.1   The Kalman Filter

The Kalman Filter is an optimal, in the least squares sense, estimator of the true state of a dynamic linear system that its measurements are corrupted by white uncorrelated noise. In our context the Kalman filter is used to estimate the vector $\hat{v}$. In its simplest form, a Kalman filter, is based on the following five equations:

$$\hat{x}_k^- = A\hat{x}_{k-1} + w_{k-1} \tag{1}$$

$$P_k^- = AP_{k-1}A^T + Q \tag{2}$$

$$K_k = P_k^- H^T (HP_k^- H^T + R)^{-1} \tag{3}$$

$$\hat{x}_k = \hat{x}_k^- + K_k(z_k - H\hat{x}_k^-) \tag{4}$$

$$P_k = (I - K_k H)P_k^- \tag{5}$$

Where $\hat{x}_k^-$ is the a priori state estimate, $\hat{x}_k$ is the a posteriori state estimate, A is the state transition matrix, w is the white, zero-mean, uncorrelated noise, P_k^- is the a priori error covariance, P_k is the a posteriori error covariance, Q is the process error covariance, R is the measurement noise covariance, H is the measurement matrix, z_k is the measurement taken at time k and K is the Kalman Gain. Equations 1 and 2 are the *Time Update* (Predict) equations. Equations 3 to 5 are the *Measurement Update* (Update or Correct) equations.

The goal of the Kalman filter is to formulate an *a posteriori* estimate $\hat{x}_k$ as a linear combination of an *a priori* estimate $\hat{x}_k^-$ and a weighted difference between an actual measurement z_k and a measurement prediction $H\hat{x}_k^-$ as shown in equation 4 [20].

Due to space limitations we will not discuss the particulars of the Kalman Filter in more detail; the interested reader can refer to [17], [20], [22] for an extensive review.

2.2   In-Network Estimation

Our localisation algorithm makes appropriate use of a Kalman filter on the basis of the assumptions of section 1.1. The process starts at an individual node — this can be a choice from a pre-determined set of nodes that sense the event independently. The desired outcome is to start a "walk" of the sensor field by visiting (i.e. sending a packet to) other nodes.

Once the event is sensed, the initiating node makes an initial guess of the measurement at adjacent single-hop nodes. Since no other information is available, this guess is equal to a linear transformation of the local measurement. The node then tasks the one-hop unvisited neighbours to send their measurements. This is achieved by a local broadcast. We will see later how the local broadcast is avoided once nodes reach a consensus regarding the quadrant direction of the source. Once the replies with the measurements have been collected, the innovations $Z_n = (z_k^{(i)} - H\hat{x}_k^-)$ are calculated. Note that $\hat{x}^-$ is the initial estimate made by the initiating node and measurement $z_k^{(i)}$ has now a superscript i to indicate which neighbouring node has reported it. Then the minimum innovation $\operatorname*{argmin}_z(z_k^{(i)} - H\hat{x}_k^-)$ is calculated and node i is selected as the next hop.

When node selection takes place, the selector sends to the selectee a collection of filter parameters so that the latter can continue the process. This allows each node in the network to run a lightweight application and only when necessary to be tasked to perform the estimation. We will refer to this collection of parameters as *particle*, since it resembles a travelling particle that facilitates a task.

The typical stopping condition is linked to the estimation relative error (line 13); for instance, when the estimation relative error exceeds a multiple of the mean relative error, the process halts. A sharp rise in the relative error usually reveals that the particle has moved outside the plume or to an area where measurements differ significantly given the initial estimates.

Moreover, a variation of the algorithm has been developed that when the relative estimation error becomes high then a selector node considers candidate next-hop nodes that are more than one hop away. The same local broadcast

---

Algorithm 1. Particle Localisation Algorithm

1: **variables** Estimate Error Covariance P, Measurement Noise Variance R, Process Variance Q, State Transition Matrix A, Measurement Matrix H, Initial Estimate $\hat{x}_k{}^-$, maxhopcount=1, netpath[], counter $c = 0$;
2: Project state estimate $\hat{x}_k{}^-$ ahead (Eq. 1).
3: Project error covariance $P_k{}^-$ ahead (Eq. 2).
4: Task *unvisited* neighbours within maxhopcount to report measurement.
5: **for** (each of replies received) **do**
6:     calculate innovations $(z_k{}^{(i)} - H\hat{x}_k{}^-)$
7: **end for**
8: Select as next hop the node that minimises the innovation.
9: Compute the Kalman gain (Eq. 3).
10: Correct (Update) estimate with measurement $z_k{}^{(i)}$ (Eq. 4).
11: Correct (Update) the error covariance P_k (Eq. 5).
12: Compute relative error.
13: **if** abs(relative error) $>=$multiple·($\mathbb{E}[Rel\ Error]$) **then**
14:     **exit**
15: **else**
16:     Add local address to netpath[c] and increment c.
17:     Send particle to selected node (line 8) and task it to start at Line 1.
18: **end if**

---

is issued but the maximum hopcount is increased to a pre-determined value i.e. to less than 3. This effectively means that the selector will receive more measurement replies — to be precise it will receive $(2h + 1)^2 - 1$ where h is the number of hops. Since there are more measurements available it is more likely to find a candidate for next hop that will keep the error low.

The typical use of this localisation algorithm is to employ many particles for robustness. With n particles we add a geometric computation — not shown in the above listing for the sake of simplicity — to establish a *consensus* regarding the quadrant in which the source is located. This operates as follows: after a small number of hops (i.e. less than 10), a convex hull is evaluated for the particles' coordinates. Recall that the convex hull is the boundary of the minimal convex set containing a finite set of points (i.e. the N particle coordinates). Provided that particles commence at nearby locations (i.e. less than 4 hops away from each other) the convex hull can be evaluated without a significant communication overhead. The mean direction of movement is given by calculating the centroid of the convex hull Fig. 3(a). Only a coarse *quadrant direction* is necessary at this point. This is the quadrant in which N particles estimate the source location and it constitutes the consensus. Once the consensus becomes known, there is no need for local broadcasts. Instead, the estimates are sent directly to the neighbours in the consensus direction. This achieves an improvement both in terms of communication — $\frac{3}{4}$ less messages at each hop — and estimate accuracy.

3   Evaluation

Given that empirical evaluation using realistic conditions is problematic due to the hazardous application scenarios, we have evaluated the correctness of the algorithm via simulations in MATLAB. We have considered a specific test case of a time-averaged gas plume over a 100-by-100 square grid according to the model described in [7] and the assumptions of section 1.1. The performance criteria used were: (a). the Euclidean distance to the true source and (b). the length of the path taken from the start to the end point that comprised the final estimate. The former criterion is associated with estimation accuracy while the latter is a measure of speed, communication cost and efficiency. Moreover we have employed a naive algorithm that selects the maximum reading as the next hop as a baseline benchmark.

Fig. 1 exhibits the simplest case involving a single particle starting at an arbitrary location and iteratively moving towards the source region of the spatial event until it reaches at $(93, 93)$, a distance of 2.83 away from the true source — a useful estimate of the true coordinates and an intensity estimate within $\epsilon = .0121$ of the true intensity. Taking into account that network-wide broadcasts were not required and only local communication between a very small percentage of the nodes was involved, we assert that the method is lightweight and respects the constraints of the devices. In addition, although the starting coordinates can be any point in two-dimensional space, for the sake of demonstrating one of the strengths of the algorithm, in this discussion we will assume starting points located fairly far from the source.

Fig. 2 shows the benchmark comparison to the naive algorithm; this small set of random starting points suggests correct and efficient behaviour. The fact that all paths taken by the 8 particles achieved a distance less than 4 to the true source in

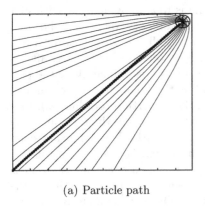

(a) Particle path

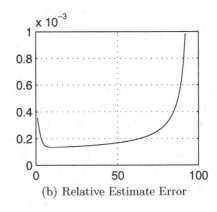

(b) Relative Estimate Error

Fig. 1. An example of the particle path (a) shown by the darker line. Starting at $(1, 1)$ it moves towards the source at $(95, 95)$. It reaches a distance of 2.83 away after 93 hops. On (b) the estimation error is shown. While, initially it is reducing as the particle approaches the source it increases gradually and then sharply — the sharp rise indicates that the measurements are not in line with expectations and it is the stopping condition.

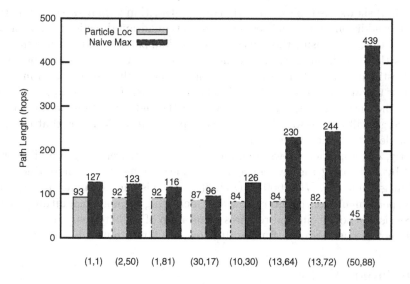

Fig. 2. Comparison of our particle localisation algorithm with a maximum selection naive algorithm. The x-axis is labelled with starting coordinates. All paths ended at a distance less than 4 (Euclidean) to the real source, but our approach resulted in shorter (in the case of $(50, 88)$, up to a factor of 9) paths.

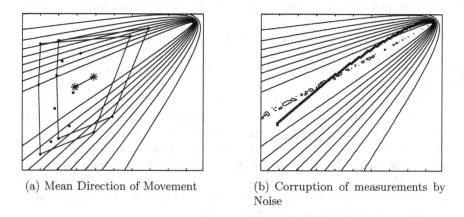

(a) Mean Direction of Movement

(b) Corruption of measurements by Noise

Fig. 3. (a) Two convex hulls evaluated for 8 particles after 10 and 20 hops respectively. Marked with an asterisk are the centroids of the polygons (consensus). (b) Particle starting at $(10, 30)$ and heading towards the source — measurements are corrupted by random noise.

less than 100 hops seems to suggest quick convergence to appropriate estimates. Furthermore, in various test runs our algorithm did not seem to be affected by the local maxima that caused the naive maximum selection algorithm to move in circular fashion and result in long paths.

Fig. 3(a) shows the typical operation of the algorithm with n particles and the geometric addition that computes the consensus quadrant direction of movement shown by the arrow. Using the consensus, individual particles only send estimates towards that quadrant direction reducing communication and achieving more robust estimates. By adding this geometric aspect to the stochastic estimation nature of the algorithm, we factor a level of error tolerance in the approach. As long as the majority of the particles move in the right direction, we guarantee that energy will not be wasted in erroneous propagation decisions that can not possibly lead to an accurate estimate.

Lastly, our approach shows tolerance to small measurement noise. This was tested by corrupting the gas concentration measurements by random noise of small magnitude (i.e. $N(0, .0001)$) (Fig. 3(b)) — this had no impact to the estimation performance of the filter. However, increasing the noise magnitude results in inconsistent behaviour and requires adaptation of the filter parameters which we will not discuss here any further due to space limitations.

4   Related Work

Coarse grained localisation techniques can be as simple as the *Centroid Calculation* or Point-in-Triangle (PIT) methods [9]. A refinement of the PIT technique is the Approximate Point in Triangle (APIT) described in [6]. A geometric approach based on the circles of Apollonius is described in [4]. A robust to noise and measurement errors data fusion algorithm that extends the latter approach is presented in [2].

A different family of methods is based on the well-understood Time Difference of Arrival (TDoA) localisation. There is a geometric and a numerical solution to this problem — more information on the details of TDoA solutions can be found in [11], [13], and [21].

Another solution is based on Maximum Likelihood Estimation (MLE): the least squares minimisation of the estimate of vector $\hat{v}$ can be solved by gradient descent or numerically according to the methodologies of [5] and [14].

Lastly, the Kalman filter has been employed in radioactive source localisation in [5]. The difference with our approach is that we perform the estimation in-the-network while [5] assumes offline processing. A combination of a Kalman filter with Time of Arrival (ToA) is presented in [8] while a distributed Kalman filter for wireless sensor networks is presented in [18].

5   Future Work and Discussion

To validate the correctness of the simulation results, at the time of this writing, we are in the process of implementing the algorithm in a network-specific simulator [1]. This simulator accepts TinyOS code and allows evaluating the impact of many real-world network-specific parameters such as density, distribution and so on.

To conclude, we have presented an iterative point-source coarse localisation algorithm that operates in-the-network and does not require powerful nodes or network-wide collection and offline processing. A straightforward Kalman filter is at the heart of the algorithm that iteratively computes the source estimate. Using this approach has the following advantages:

- *Efficient paths* in terms of length and distance to the source, when compared to a naive maximum selection algorithm.
- *Robustness* which is introduced by using multiple particles and a geometric approach to establish mean direction of movement. This direction consensus facilitates both reduction in communication and improved accuracy of estimates.
- *Lightweight*, real-time properties that neither involve the entire network nor make any assumptions about individual placement of nodes.
- *Noise resilience* due to intrinsic characteristics of the Kalman filter make our approach robust to errors introduced by the inexpensive circuitry and sensory equipment.

Finally, the recursive nature of the Kalman filter makes it a good match for operation in resource-constrained devices — approaches such as [19] use optimised implementations — and a valid generic solution to the CBR source localisation problem.

References

1. Avrora - The AVR Simulation and Analysis Framework,
   http://compilers.cs.ucla.edu/avrora
2. Chin, J.-C., Yau, D.K.Y., Rao, N.S.V., Yang, Y., Ma, C.Y.T., Shankar, M.: Accurate localization of low-level radioactive source under noise and measurement errors. In: SenSys 2008: Proceedings of the 6th ACM conference on Embedded network sensor systems, New York, NY, USA, pp. 183–196 (2008)
3. Sentilla Corp. TMote Sky Datasheet,
   http://www.sentilla.com/pdf/eol/tmote-sky-datasheet.pdf
4. Cox, J., Partensky, M.: Spatial Localization Problem and the Circle of Apollonius (January 2007)
5. Gunatilaka, A., Ristic, B., Gailis, R.: Radiological Source Localisation (DSTO-TR-1988) (1988)
6. He, T., Huang, C., Blum, B.M., Stankovic, J.A., Abdelzaher, T.: Range-free localization schemes for large scale sensor networks. In: MobiCom 2003: Proceedings of the 9th annual international conference on Mobile computing and networking, New York, USA, pp. 81–95 (2003)
7. Ishida, H., Nakamoto, T., Moriizumi, T.: Remote sensing of gas/odor source location and concentration distribution using mobile system. Solid State Sensors and Actuators 1(16), 559–562 (1997)
8. Klee, U., Gehrig, T., McDonough, J.: Kalman filters for time delay of arrival-based source localization. EURASIP J. Appl. Signal Process. 1, 167 (2006)
9. Krishnamachari, B.: Networking Wireless Sensors. Cambridge University Press, Cambridge (2005)
10. Levis, P., Patel, N., Culler, D., Shenker, S.: Trickle: a self-regulating algorithm for code propagation and maintenance in wireless sensor networks. In: NSDI 2004: Proceedings of the 1st conference on Symposium on Networked Systems Design and Implementation, Berkeley, CA, USA (2004)

11. Mellen, G., Pachter, M., Raquet, J.: Closed-form solution for determining emitter location using time difference of arrival measurements. IEEE Transactions on Aerospace and Electronic Systems 39(3), 1056–1058 (2003)
12. BBC News Online. Threat of dirty bombs 'increased' (March 2009), http://news.bbc.co.uk/1/hi/uk/7960466.stm
13. Rao, N.: Identification of simple product-form plumes using networks of sensors with random errors. In: 9th International Conference on Information Fusion, pp. 1–8 (2006)
14. Savvides, A., Han, C.-C., Strivastava, M.B.: Dynamic fine-grained localization in Ad-Hoc networks of sensors. In: MobiCom 2001: Proceedings of the 7th annual international conference on Mobile computing and networking, New York, USA, pp. 166–179 (2001)
15. Home Office Office for Security and Counter Terrorism. Chemical, Biological, Radiological, Nuclear Resilience, http://security.homeoffice.gov.uk/cbrn-resillience/
16. SensorNet. Nationwide Detection and Assessment of Chemical, Biological, Nuclear and Explosive (CBRNE) Threats, http://www.sensornet.gov/sn_overview.html
17. Simon, D.: Optimal State Estimation: Kalman, H Infinity, and Nonlinear Approaches. Wiley Blackwell, Chichester (2006)
18. Spanos, D.P., Olfati-Saber, R., Murray, R.M.: Approximate distributed Kalman filtering in sensor networks with quantifiable performance. In: IPSN 2005: Proceedings of the 4th international symposium on Information processing in sensor networks, Piscataway, NJ, USA, p. 18 (2005)
19. Tan, J., Kyfuakopoulos, N.: Implementation of a Tracking Kalman Filter on a Digital Signal Processor. IEEE Transactions on Industrial Electronics 35(1), 126–134 (1988)
20. Welch, G., Bishop, G.: An Introduction to the Kalman Filter. Technical Report 95-041, Chapel Hill, NC, USA (1995)
21. Xu, X., Rao, N., Sahni, S.: A computational geometry method for DTOA triangulation. In: 10th International Conference on Information Fusion, pp. 1–7 (2007)
22. Zarchan, P.: Fundamentals of Kalman Filtering: A Practical Approach, 2nd edn. Progress in Astronautics & Aeronautics Series. AIAA (2005)
23. Zoumboulakis, M., Roussos, G.: Escalation: Complex event detection in wireless sensor networks. In: Kortuem, G., Finney, J., Lea, R., Sundramoorthy, V. (eds.) EuroSSC 2007. LNCS, vol. 4793, pp. 270–285. Springer, Heidelberg (2007)

Hyperellipsoidal SVM-Based Outlier Detection Technique for Geosensor Networks

Yang Zhang, Nirvana Meratnia, and Paul Havinga

Pervasive Systems Group, University of Twente,
Drienerlolaan 5, 7522NB Enschede, The Netherlands
{zhangy,meratnia,havinga}@cs.utwente.nl

Abstract. Recently, wireless sensor networks providing fine-grained spatiotemporal observations have become one of the major monitoring platforms for geo-applications. Along side data acquisition, outlier detection is essential in geosensor networks to ensure data quality, secure monitoring and reliable detection of interesting and critical events. A key challenge for outlier detection in these geosensor networks is accurate identification of outliers in a distributed and online manner while maintaining low resource consumption. In this paper, we propose an online outlier detection technique based on one-class hyperellipsoidal SVM and take advantage of spatial and temporal correlations that exist between sensor data to cooperatively identify outliers. Experiments with both synthetic and real data show that our online outlier detection technique achieves better detection accuracy compared to the existing SVM-based outlier detection techniques designed for sensor networks. We also show that understanding data distribution and correlations among sensor data is essential to select the most suitable outlier detection technique.

Keywords: Geosensor networks, outlier detection, data mining, one-class support vector machine, spatio-temporal correlation.

1 Introduction

Advances in sensor technology and wireless communication have enabled deployment of low-cost and low-power sensor nodes that are integrated with sensing, processing, and wireless communication capabilities. A geosensor network consists of a large number of these sensor nodes distributed in a large area to collaboratively monitor phenomena of interest. The monitored geographic space may vary in size and can range from small-scale room-sized spaces to highly complex dynamics of ecosystem regions [1]. Emerging applications of large-scale geosensor networks include environmental monitoring, precision agriculture, disaster management, early warning systems and wildlife tracking [1]. In a typical application, a geosensor network collects and analyzes continuous streams of fine-grained geosensor data, detects events, makes decisions, and takes actions.

Wireless geosensor networks have strong resource constraints in terms of energy, memory, computational capacity, and communication bandwidth. Moreover, the

N. Trigoni, A. Markham, and S. Nawaz (Eds.): GSN 2009, LNCS 5659, pp. 31–41, 2009.

autonomous and self-organizing vision of these networks makes them a good candidate for operating in harsh and unattended environments. Resource constraints and environmental effects cause wireless geosensor networks to be more vulnerable to noise, faults, and malicious activities (e.g., denial of service attacks or black hole attacks), and more often generate unreliable and inaccurate sensor readings. Thus, to ensure a reasonable data quality, secure monitoring and reliable detection of interesting and critical events, identifying anomalous measurements locally at the point of action (at the sensor node itself) is a must. These anomalous measurements, usually known as outliers or anomalies, are defined as measurements that do not conform with the normal behavioral pattern of the sensed data [2].

Unlike traditional outlier detection techniques performed off-line in a centralized manner, limited resources available in sensor networks and specific nature of geosensor data necessitate outlier detection to be performed in a distributed and online manner to reduce communication overhead and enable fast responce. This implies that outliers in distributed streaming data should accurately be detected in real-time while maintaining resource consumption low. In this paper, we propose an online outlier detection technique based on one-class hyperellipsoidal Support Vector Machine (SVM) and take advantage of spatial and temporal correlation that exist between sensor data to cooperatively identify outliers. Experiments with both synthetic and real data obtained from the EPFL SensorScope System [3] show that our online outlier detection technique achieves better detection accuracy and lower false alarm compared to the existing SVM-based outlier detection techniques [4], [5] designed for sensor networks.

The remainder of this paper is organized as follows. Related work on one-class SVM-based outlier detection techniques is presented in Section 2. Fundamentals of the one-class hyperellipsoidal SVM are described in Section 3. Our proposed distributed and online outlier detection technique is explained in Section 4. Experimental results and performance evaluation are reported in Section 5. The paper is concluded in Section 6 with plans for future research.

2   Related Work

Generally speaking, outlier detection techniques can be classified into statistical-based, nearest neighbor-based, clustering-based, classification-based, and spectral decomposition-based approaches [2], [6]. SVM-based techniques are one of the popular classification-based approaches due to the fact that they (i) do not require an explicit statistical model, (ii) provide an optimum solution for classification by maximizing the margin of the decision boundary, and (iii) avoid the curse of dimensionality problem.

One of the challenges faced by SVM-based outlier detection techniques for sensor networks is obtaining error-free or labelled data for training. One-class (unsupervised) SVM-based techniques can address this challenge by modelling the normal behavior of the unlabelled data while automatically ignoring the anomalies in the training set. The main idea of one-class SVM-based outlier detection techniques is to use a non-linear function to map the data vectors (measurements) collected from the original space (input space) to a higher dimensional space (feature space). Then a decision boundary of normal data will be found that encompasses the majority of the data vectors in the feature space. Those new unseen data vectors falling outside the boundary are classified as outliers. Scholkopf et al. [7] have proposed a hyperplane-based one-class SVM, which identifies outliers by fitting a hyperplane from the origin. Tax et al. [8] have proposed a hypersphere-based one-class SVM, which identifies outliers by fitting a hypersphere with a minimal radius. Wang et al. [9] have proposed a hyperellipsoid-based one-class SVM, which identifies outliers by fitting multiple hyperellipsoids with minimum effective radii.

In addition to obtaining the labelled data, another challenge faced by SVM-based outlier detection techniques is their quadratic optimization during the learning process for the normal boundary. This process is extremely costly and not suitable for limited resources available in sensor networks. Laskov et al. [10] have extended work in [8] by proposing a one-class quarter-sphere SVM, which is formulated as a linear optimization problem by fitting a hypersphere centered at the origin and thus reducing the effort and computational complexity. Rajasegarar et al. [11] and Zhang et al. [5] have further exploited potential of the one-class quarter-sphere SVM of [10] for distributed outlier detection in sensor networks. The main difference of these two techniques is that unlike a batch technique of [11], the work of [5] aims at identifying every new measurement collected at a node as normal or anomalous in an online manner.

Rajasegarar et al. [4] have also extended work in [9] [10] by proposing a one-class centered hyperellipsodal SVM with linear optimization. However, this technique is neither distributed nor online. In this paper, we extend work in [4] and propose a distributed and online outlier detection technique suitable for geosensor networks, with low computational complexity and memory usage.

3 Fundamentals of the One-Class Hyperellipsoidal SVM

In our proposed technique, we exploit the one-class hyperellipsoidal SVM [9], [4] to learn the normal behavioral pattern of sensor measurements. The quadric optimization problem of the one-class hyperellipsoidal SVM has been converted to a linear optimization problem in [4] by fixing the center of the hyperellipsoidal at the origin. A hyperellipsoidal boundary is used to enclose the majority of the data vectors in the feature space. The geometries of the one-class centered hyperellipsoidal SVM-based approach is shown in Fig. 1.

The constrain for optimization problem of the one-class centered hyperspherical SVM is formalized as follows:

$$\min_{R\epsilon\Re, \xi\epsilon\Re^m} R^2 + \frac{1}{\upsilon m} \sum_{i=1}^{m} \xi_i \tag{1}$$

$$subject\ to: \phi(x_i)\Sigma^{-1}\phi(x_i)^T \leq R^2 + \xi_i, \xi_i \geq 0, i = 1, 2, \dots m$$

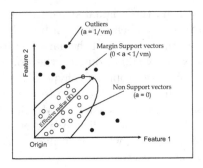

Fig. 1. Geometry of the hyperellipsoidal formulation of one-class SVM [4]

where m denotes number of data vectors in the training set. The parameter $\upsilon \,\epsilon\, (0, 1)$ controls the fraction of data vectors that can be outliers. Σ^{-1} is the inverse of the covariance matrix $\Sigma = \frac{1}{m} \sum_{i=1}^{m} (\phi(x_i) - \mu)(\phi(x_i) - \mu)^T$, $\mu = \frac{1}{m} \sum_{i=1}^{m} \phi(x_i)$. Using Mercer Kernels [12], the dot product computations of data vectors in the feature space can be computed in the input data space. The centered kernel matrix K_c can be obtained in terms of the kernel matrix K using $K_c = K - 1_m K - K 1_m + 1_m K 1_m$, where 1_m is the $m \times m$ matrix with all values equal to $\frac{1}{m}$. Finally, the dual formulation of (1) will become a linear optimization problem formulated as follows:

$$\min_{\alpha \epsilon \Re^m} - \sum_{i=1}^{m} \alpha_i \| \sqrt{m} \Lambda^{-1} P^T K_c^i \|^2 \tag{2}$$

$$subject \ to: \sum_{i=1}^{m} \alpha_i = 1, 0 \leq \alpha_i \leq \frac{1}{\upsilon m}, i = 1, 2, \ldots m$$

where Λ is a diagonal matrix with positive eigenvalues, P is the eigenvector matrix corresponding to the positive eigenvalues [13], and K_c^i is the ith column of the kernel matrix K_c. From equation (2), the $\{\alpha_i\}$ value can be easily obtained using some effective linear optimization techniques [14]. The data vectors in the training set can be classified depending on the results of $\{\alpha_i\}$, as shown in Fig. 1. The training data vectors with $0 \leq \alpha \leq \frac{1}{\upsilon m}$, which fall on the hyperellipsoid, are called margin support vectors. The effective radius of the hyperellipsoid $R = \| \sqrt{m} \Lambda^{-1} P^T K_c^i \|$ can be computed using any margin support vector.

4   A Distributed and Online Outlier Detection Technique for GeoSensor Networks

In this section, we will describe our distributed and online outlier detection technique. This proposed technique aims at identifying every new measurement collected at each node as normal or anomalous in real-time. Moreover, using high degree of spatio-temporal correlations that exist among the sensor readings, each node exchanges the learned normal boundary with its neighboring nodes and combines their learned normal boundaries to cooperatively identify outliers. Before describing this technique in detail, we present our assumptions and explain why we exploit the hyperellipsoidal SVM instead of hyperspherical SVM to learn the normal behavioral pattern of sensor measurements.

4.1 Assumptions

We assume that wireless sensor nodes are time synchronized and densely deployed in a homogeneous geosensor network, where sensor data tends to be correlated in both time and space. A sensor sub-network consists of n sensor nodes $S_1, S_2, \ldots S_n$, which are within radio transmission range of each other. This means that each node has $n\text{-}1$ neighboring nodes in the sub-network. At each time interval Δ_i, each sensor node in the sub-network measures a data vector. Let $x_1^i, x_2^i, \ldots, x_n^i$ denote the data vector measured at $S_1, S_2, \ldots S_n$, respectively. Each data vector is composed of multiple attributes x_j^{il}, where $x_j^i = \{x_j^{il} : j = 1 \ldots n, l = 1 \ldots d\}$ and $x_j^i \in \Re^d$. Our aim is online identification of every new measurement collected at each node as normal or anomalous by means of local processing at the node itself. In addition to near real-time identification of outliers, increasing data quality, and reducing communication overhead, this local processing also has the advantage of coping with (possibly) large scale of the geosensor network.

4.2 Hyperellipsoidal SVM vs. Hyperspherical SVM

In this paper, we exploit the hyperellipsoidal SVM instead of hyperspherical SVM to learn the normal behavioral pattern of sensor measurements. The reason for doing so is the fact that hyperspherical SVM assumes that the target sample points are distributed around the center of mass in an ideal spherical manner. However, if the data distribution is non-spherical, using a spherical boundary to fit the data will increase the false alarm rate and reduce the detection rate. This is because many superfluous outlier are mistakenly considered in the boundary and consequently outliers are classified as normal.

On the contrary, the hyperellipsoidal SVM is able to best capture multivariate data structures by considering not only the distance from the center of mass but also the data distribution trend, where the latter is learned by building the covariance matrix of the sample points. This feature can be used well for geosensor data, where multivariate attributes may induce certain correlation, e.g., the readings of humidity sensors are negatively correlated to the readings of temperature sensors. Unlike the Euclidean distance used in the hyperspherical SVM, the distance metric adopted in the hyperellipsoidal SVM is the Mahalanobis distance. The Mahalanobis distance takes the shape of the multivariate data distribution into account and identifies the correlations of data attributes. Thus using an ellipsoidal boundary to enclose geosensor data aims to increase outlier detection accuracy and reduce the false alarm rate. However, as a tradeoff, the hyperellipsoidal SVM has more computational and memory usage cost than the hyperspherical SVM. To correctly select the most appropriate outlier detection technique, we believe that having some understanding of the data distribution and correlation among sensor data is crucial.

4.3   Hyperellipsoidal SVM-Based Outlier Detection Techniques

The main idea behind our proposed Hyperellipsoidal SVM-based online outlier detection technique (OOD_E) is that each node builds a normal boundary representing normal behavior of the sensed data and then exchanges the learned normal boundary with its spatially neighboring nodes. A sensor measurement collected at a node is identified as an outlier if it does not fit inside the boundary defined at the node and also does not fit inside the combined boundaries of the spatially neighboring nodes. We first explain the OOD_E technique in the input and feature spaces and then present the corresponding pseudocode in Table 1.

OOD_E in the Input Space. Initially, each node learns the local effective radius of the hyperellipsoid using its m sequential data measurements, which may include some anomalous data. In the input space, equation (1) can be formalized as equation (3). The one-class hyperellipsoidal SVM can efficiently find a minimum effective radius R to enclose the majority of these sensor measurements in the input space. Each node then locally broadcasts the learned radius information to its neighboring nodes. When receiving the radii from all of its neighbors, each node computes a median radius R_m of its neighboring nodes. We use median because in estimating the "center" of a sample set, the median is more accurate than the mean.

$$\min_{R \in \Re, \xi \in \Re^m} R^2 + \frac{1}{vm} \sum_{i=1}^{m} \xi_i \qquad (3)$$

$$subject\ to : (x_i - \mu)\Sigma^{-1}(x_i - \mu)^T \leq R^2 + \xi_i, \xi_i \geq 0, i = 1, 2, \ldots m$$

Sensor data collected in a densely deployed geosensor network tends to be spatially and temporally correlated [1]. When a new sensor measurement x is collected at node S_i, node S_i first compares the Mahalanobis distance of x with its local effective radius R_i. In the input space, the mean can be expressed as $\mu = \frac{1}{m} \sum_{i=1}^{m} x_i$, and thus the Mahalanobis distance of x is formulated as follows:

$$Md(x) = \sqrt{(x - \mu)\Sigma^{-1}(x - \mu)^T} = \|\Sigma^{-\frac{1}{2}}(x - \frac{1}{m}\sum_{i=1}^{m} x_i)\| \qquad (4)$$

The data x will be classified as normal if $Md(x) <= R_i$. This means that x falls on or inside the hyperellipsoid defined at S_i. If $Md(x) > R_i$, S_i further compares $Md(x)$ with the median radius R_{im} of its spatially neighboring nodes. Then if

$Md(x) > R_{im}$, x will finally be classified as an outlier. The decision function to declare a measurement as normal or outlier can be formulated as equation (5), where a reading with a negative value is classified as an outlier.

$$f(x) = sgn(max(R - Md(x), R_m - Md(x)))$$ (5)

The computational complexity of OOD_E in the input space is low as it only depends on solving a linear optimization problem presented in equation (3) and simple computations expressed by equations (4) and (5). Once the optimization is solved, each node only keeps the effective radius value, the mean, and the covariance matrix obtained from the training data in memory. Using the radius information from adjacent nodes is to reduce high false alarm caused by unsupervised learning techniques.

OOD_E in the Feature Space. Each node learns the local effective radius of the hyperellipsoid using equation (1) and (2), and then exchanges the learned radius information with its spatially neighboring nodes. In the input space, the mean can be expressed as $\mu = \frac{1}{m} \sum_{i=1}^{m} \phi(x_i)$, and thus the Mahalanobis distance of each new measurement x in the feature space can be formalized as follows:

$$Md(x) = \sqrt{(\phi(x) - \mu)\Sigma^{-1}(\phi(x) - \mu)^T} = \|\sqrt{m}\Lambda^{-1}P^T K_c^x\|$$ (6)

Then the data x will be classified as normal or anomalous using the same decision function as the equation (5). Due to high computational cost and memory usage required for classification of each new sensor measurement as normal or anomalous, we modify the OOD_E in such a way that it does not run for every new sensor reading and it waits until a few measurements $X = \{x^n : n = 1 \ldots t\}$ are collected. The centered kernel matrix K_c^X can be obtained by using $K_c^X = K^X - 1_{tm}K - K^X 1_m + 1_{tm}K 1_m$, where 1_{tm} is the $t \times m$ matrix with all values equal to $\frac{1}{m}$. This modification reduces the computational complexity and also facilitates linking outliers to actual events in later stages.

5   Experimental Results and Evaluation

The goals of expreximents are two folds. First we evaluate performance of our distributed and online technique compared to the batch hyperellipsoidal SVM-based outlier detection technique (BOD_E) presented in [4] and the online quarter-sphere SVM-based outlier detection technique (OOD_Q) presented in [5]. Secondly, we investigate the impact of data distribution and spatial/spatio-temporal correlations in performance of our outlier detection technique. In experiments, we use synthetic data as well as real data gathered from a geosensor network deployment by the EPFL [3].

Table 1. Pseudocode of the OOD_E

---

1 procedure **LearningSVM()**
2   each node collects m sensor measurements for learning its own effective radius R
     and locally broadcasts the radius to its spatially neighboring nodes;
3   each node then computes R_m;
4   initiate **OutlierDetectionProcess**(R, R_m);
5   return;

6 procedure **OutlierDetectionProcess**(R, R_m)
7   when a new measurement x arrives
8   compute $Md(x)$;
9   if $(Md(x) > R$ AND $Md(x) > R_m)$
10   x indicates an outlier;
11   else
12   x indicates a normal measurement;
13   endif;
14   return;

5.1 Datasets

For the simulation, we use Matlab and consider a sensor sub-network consisting of seven sensor nodes. Sensor nodes are within the one-hop range of each other. Two 2-D synthetic data distributions with 10% (of the normal data) anomalous data are shown in Fig. 2(a) and 4(a). It can clearly be seen that Fig. 4(a) has a concentrated distribution around the origin while the data distribution shown in Fig. 2(a) is not spherical but has a certain trend. The data values are normalized to fit in the [0, 1]. The BOD_E performs outlier detection when all measurements are collected at each node, while the OOD_Q operates in a distributed and online manner.

The real data is collected from a closed neighborhood by a geosensor network deployed in Grand-St-Bernard. Fig. 3(a) shows the deployment area. The closed neighborhood contains the node 31 and its 4 spatially neighboring nodes, namely nodes 25, 28, 29, 32. The network records ambient temperature, relative humidity, soil moisture, solar radiation and watermark measurements at 2 minutes intervals. In our experiments, we use ambient temperature and relative humidity collected during the period of 9am-5pm on the 5th October 2007. The labels of measurements are obtained based on degree of dissimilarity between data measurements.

5.2 Experimental Results and Evaluation

We have evaluated two important performance metrics, the detection rate (DR), which represents the percentage of anomalous data that are correctly classified as outliers, and the false alarm rate, also known as false positive rate (FPR), which represents the percentage of normal data that are incorrectly considered as outliers. A receiver operating characteristics (ROC) curve is used to represent the trade-off between the detection rate and the false alarm rate. The larger the area under the ROC curve, the better the performance of the technique.

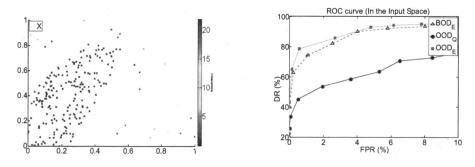

Fig. 2. (a) Plot for synthetic data; (b) ROC curves in the input space

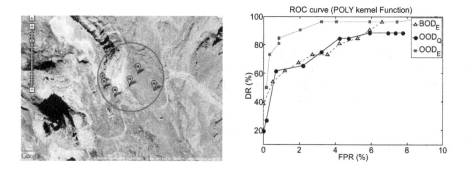

Fig. 3. (a) Grand-St-Bernard deployment in [3]; (b) ROC curves in the feature space

We have examined the effect of the regularization parameter v for OOD_E, BOD_E, and OOD_Q. v represents the fraction of data vectors that can be outliers. For synthetic dataset, we varied v from 0.02 to 0.18 in intervals of 0.02 and evaluated the detection accuracy of the three techniques in the input space. For real dataset, we varied v from 0.01 to 0.10 in intervals of 0.01 and used Polynomial kernel function to evaluate the accuracy performance of three techniques in the feature space. The Polynomial kernel function is formulated as: $k_{POLY} = (x_1.x_2 + 1)^r$, where r is the degree of the polynomial.

Fig. 2(b) and 3(b) show the ROC curves obtained for the three techniques in the input space for synthetic data and using Polynomial kernel function for real data. Simulation results show that our OOD_E always outperforms BOD_E and OOD_Q. Moreover, quarter-sphere SVM-based OOD_Q is obviously worse than the hyperellipsoidal SVM-based OOD_E and BOD_E in the input space for synthetic data. For real data in the feature space, the performance of OOD_Q and BOD_E is not very obvious to distinguish.

Although experiments show that hyperellipsoidal SVM-based techniques outperform quarter-sphere SVM-based technique, our experiments show that this greatly depends on data distribution and correlations that exist between sensor data. It can be clearly seen from Fig. 4(b), the quarter-sphere SVM-based

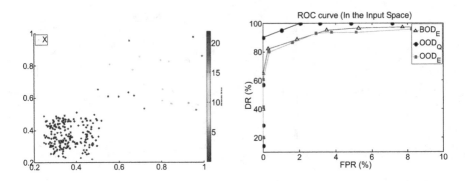

Fig. 4. (a) Plot for synthetic data; (b) ROC curves in the input space

OOD_Q has a better performance than two hyperellipsoidal SVM-based outlier detection techniques in the input space for synthetic data. The obtained results conform with our idea about the need of having some understanding of the data distribution and correlation among sensor data to be able to select the most suitable outlier detection technique.

6   Conclusions

In this paper, we have proposed a distributed and online outlier detection technique based on one-class hyperellipsoidal SVM for geosensor networks. We compare the performance of our techniques with the existing SVM-based techniques using both synthetic and real data sets. Experimental results show that our technique achieves better detection accuracy and lower false alarm. Our future research includes sequentially updating the normal boundary of the sensor data, online computation of spatio-temporal correlations and online distinction outliers between events and errors.

Acknowledgments. This work is supported by the EU's Seventh Framework Programme and the SENSEI project.

References

1. Nittel, S., Labrinidis, A., Stefanidis, A.: GeoSensor Networks. Springer, Heidelberg (2006)
2. Chandola, V., Banerjee, A., Kumar, V.: Anomaly Detection: A Survey. Technical report, University of Minnesota (2007)
3. SensorScope, http://sensorscope.epfl.ch/index.php/Main_Page
4. Rajasegarar, S., Leckie, C., Palaniswami, M.: CESVM: Centered Hyperellipsoidal Support Vector Machine Based Anomaly Detection. In: IEEE International Conference on Communications, pp. 1610–1614. IEEE Press, Beijing (2008)

5. Zhang, Y., Meratnia, N., Havinga, P.J.M.: An Online Outlier Detection Technique for Wireless Sensor Networks using Unsupervised Quarter-Sphere Support Vector Machine. In: 4th International Conference on Intelligent Sensors, Sensor Networks and Information Processing, pp. 151–156. IEEE Press, Sydney (2008)
6. Zhang, Y., Meratnia, N., Havinga, P.J.M.: Outlier Detection Techniques for Wireless Sensor Network: A Survey. Technical report, University of Twente (2008)
7. Scholkopf, B., Platt, J.C., Shawe-Taylor, J.C., Smola, A.J., Williamson, R.C.: Estimating the Support of a High-Dimensional Distribution. Journal of Neural Computation 13(7), 1443–1471 (2001)
8. Tax, D.M.J., Duin, R.P.W.: Support Vector Data Description. Journal of Machine Learning 54(1), 45–56 (2004)
9. Wang, D., Yeung, D.S., Tsang, E.C.C.: Structured One-Class Classification. IEEE Transactions on System, Man and Cybernetics 36(6), 1283–1295 (2006)
10. Laskov, P., Schafer, C., Kotenko, I.: Intrusion Detection in Unlabeled Data with Quarter Sphere Support Vector Machines. In: Detection of Intrusions and Malware & Vulnerability Assessment, pp. 71–82. Dortmund (2004)
11. Rajasegarar, S., Leckie, C., Palaniswami, M., Bezdek, J.C.: Quarter Sphere based Distributed Anomaly Detection in Wireless Sensor Networks. In: IEEE International Conference on Communications, pp. 3864–3869. IEEE Press, Glasgow (2007)
12. Vapnik, V.N.: Statistical Learning Theory. John Wiley & Sons, Chichester (1998)
13. Golub, G.H., Loan, C.F.V.: Matrix Computations. John Hopkins (1996)
14. Nash, S.G., Sofer, A.: Linear and Nonlinear Programming. McGraw-Hill, New York (1996)

Genetic Algorithm for Clustering in Wireless Adhoc Sensor Networks

Rajeev Sachdev and Kendall E. Nygard

North Dakota State University, Fargo ND 58102, USA
rajeev.sachdev@gmail.com, kendall.nygard@ndsu.edu

Abstract. Sensor networks pose a number of challenging conceptual and optimization problems. A fundamental problem in sensor networks is the clustering of the nodes into groups served by a high powered relay head, then forming a backbone among the relay heads for data transfer to the base station. We address this problem with a genetic algorithm (GA) as a search technique.

1 Introduction

Clustering techniques were originally conceived by Aristotle and Theophrastos in the fourth century B.C. and in the 18th century by Linnaeus [1]. Even the simplest clustering problems are known to be NP-Hard [2], for instance the Euclidean k-center problem in the plane is NP-Hard [3]. Genetic algorithms can provide good solutions for such optimization problems. This study concerns the development and empirical testing of a Genetic Algorithm applied to clustering of sensor nodes in a Wireless Ad hoc Sensor Network. We test the procedure on problems in which the sensor nodes are randomly distributed over a geographical area. The genetic search represents the positions of high powered relay heads (having greater battery life than the rest of the sensor nodes) as an artificial chromosome representation, and seeks a high-performance set of assignments of nodes to a pre-specified number of clusters.

Werner and Fogarty [4] devised a genetic algorithm for a clustering problem using a binary representation for encoding a chromosome. Our approach uses floating point as an encoding scheme for the chromosomes. Painho and Bacao [5] address the clustering problem with a genetic algorithm that seeks to minimize the square error of the cluster dispersion using an approach similar to k-Means. These approaches encounter scalability issues, that is, as the number of data points is increased, the algorithm fails to form clusters. Our approach resolves this scalability issue.

2 Methodology

Following the genetic algorithm paradigm, we create a population of n artificial chromosomes that represent relay point locations. Every chromosome is evaluated using a fitness function and those with the highest fitness values are selected

N. Trigoni, A. Markham, and S. Nawaz (Eds.): GSN 2009, LNCS 5659, pp. 42–50, 2009.

as parents to take part in crossover. Based upon these point locations, a Voronoi diagram is calculated which defines the clusters. In building a Voronoi diagram, the boundaries of each cluster are defined and there is no overlapping of the sensor nodes between the clusters. Every chromosome represents a different location of relay heads which leads to different boundaries every time. Once the Voronoi diagram is generated and clusters are defined, within each cluster the distances of the sensors from their relay point is calculated using a distance metric. This metric, along with an error measure, is used to define a fitness function. A Voronoi diagram decomposes the space into regions around each site.

Let $S = \{p_1, p_2, \ldots, p_i, \ldots, p_n\}$ be a set of points (sites) in Euclidean 2-space. Using the definition in [6], the Voronoi region, $V(p_i)$, for each p_i is given by

$$V(p_i) = \{x : |p_i - x| \leq |p_j - x|, \, for \, all \, j \neq i\} \tag{1}$$

$V(p_i)$ consists of all points that are closer to p_i than any other site. The set of all sites forms the Voronoi Diagram $V(S)$ [7].

The pseudocode of our genetic algorithm is given below.

2.1   Psuedocode

```
START
 Generate random initial population
 REPEAT
    FOR every chromosome in the population
       Generate Voronoi boundaries
       FOR every cluster in chromosome
          Compute closed polygon of that region
          Compute the number of sensor nodes of that cluster/region
          Compute distance from relay head using Euclidean or MST
          Compute cluster error based on the variance factor
          Fitness = cluster error * distance
          Total chromosome fitness = sum of fitness of all Clusters
       ENDFOR
       Calculate probability of each chromosome based on fitness
       Select parents based on roulette wheel selection
       Perform Elitism by placing best chromosome in new population
       Perform one-point crossover
       Perform Mutation
    ENDFOR
    Copy the new population to old population
 UNTIL load is balanced
END
```

2.2   Chromosome Representation

Floating point numbers are used to encode the chromosomes which are initially determined randomly based on the uniform random generator. The chromosome length is equal to the pre-specified number of clusters multiplied by two (as each relay head has both x and y coordinates).

$$Ch_i^j = \{x_1^1, y_1^1, x_2^1, y_2^1, x_3^1, y_3^1, \ldots, x_n^1, y_n^1\} \tag{2}$$

In equation 1., gene i of chromosome j is an ordered pair that represents the coordinate position of a relay head in Euclidean 2-Space.

We use ρ to denote the population set, as shown in equation 2..

$$\rho = \{ch_1, ch_2, ch_3 \cdots, ch_n\} \tag{3}$$

After the initial population is generated, for every chromosome based on the relay head positions, a Voronoi diagram is generated to define the clusters, which is evaluated using the fitness function described below.

2.3   Fitness

The fitness function has two parts. The first part is the evaluation of the distance from the relay head of a particular cluster to all the sensor nodes belonging to that cluster. The distance can be provided by the Euclidean metric or a Minimum Spanning Tree. The second part is the Cluster Error which is the deviation of the number of sensor nodes per cluster from an idealised number. The fitness of a chromosome is the product of the distance and the Cluster Error. Low fitness valued are preferred as energy conservation is of high importance in Wireless Sensor Networks, since the sensors tend to be battery operated. By minimizing the distance, we conserve the energy of each sensor nodes.

The fitness function is given as follows:

$$\delta(cl) = \psi(cl) \times \xi(cl) \tag{4}$$

where,
$\delta(cl)$= Fitness of a Cluster
$\psi(cl)$= Fitness Method 1 or 2
$\xi(cl)$= Cluster Error
The lower the value of $\delta(cl)$, denser the cluster is.

$$\phi(ch) = \sum_{i=1}^{n} \delta(cl_i) \tag{5}$$

where, $\phi(ch)$ is the function to calculate the fitness of chromosome.

Fitness Function 1: Euclidean Distance. We calculate the Euclidean distance from each relay head to all the sensor nodes belonging to that particular cluster. Doing so maintains a spatial relationship between the sensor nodes and the relay heads. The shorter the distance from relay head to its respective sensor nodes, the better the cluster. Let the relay head be represented by

$$R(x, y) \tag{6}$$

The position of sensor nodes is represented by

$$S_{i=n}^{n}(x_i, y_i) \tag{7}$$

where $n =$ number of sensor nodes in that cluster.

Therefore, fitness function method 1 is defined by

$$\psi(cl) = \sum_{i=1}^{n} \sqrt{(y_i - y) + (x_i - x)} \tag{8}$$

Fitness Function 2: Minimum Spanning Tree. The second fitness method is based upon a minimum spanning tree over all of the senor nodes and the relay head. In a sensor network with relay heads, minimum spanning tree routing results in minimum cost [8]. We use the well-known Prim algorithm to compute a minimum spanning tree as the complexity of the Prim algorithm is $O(n^2)$. Moreover, the Prim algorithm is faster for dense graphs than the alternative Kruskal algorithm, which is faster for sparse graphs.[9]

Therefore,

$\psi(Cl) =$ Sum of MST distance from all sensor nodes to relay head.

2.4   Cluster Error

Recognizing that load balancing is desirable, we define a cluster error measure that is the difference between the number of nodes in a cluster as determined by the procedure and an idealized number of nodes per cluster. The idealized number d is given as follows:

$$d = \left\lfloor \left(\frac{\text{Number Of Sensor Nodes}}{\text{Number Of Clusters}} \right) \right\rfloor \tag{9}$$

Therefore,

$$\xi(cl) = |d - actual\,nodes\,per\,cluster| \tag{10}$$

This enforces a load balancing goal, in which ideally each relay head has approximately equal number of nodes in its cluster.

The larger the cluster error, for the clusters in the chromosome, the lesser the chance of it being selected as a parent for reproduction.

2.5   Selection, Crossover and Mutation

The selection process selects chromosomes from the mating pool according to the survival of the fittest concept of natural genetic system. In each successive generation, a proportion of the existing population is selected to breed a new generation. Our approach uses 80% as crossover probability, which means that 80% of the population will take part in crossover. The probabilities for each chromosome is calculated according to their fitness values, and selection is in proportion to these probabilities where the chromosome with lower probability has more chance of being selected. The proportions are calculated as given below.

$$Prob(ch_i) = \frac{fitness(ch_i)}{\sum_{i=1}^{n} fitness(ch_i)} \quad (11)$$

Once the probabilities are calculated, Roulette Wheel selection [10] is used to select parents for crossover. After the parents are selected, crossover is performed based on one-point crossover by selecting a crossover point randomly between 1 and the length of the chromosome.

We then carry out a mutation operator on all non-elite chromosomes to prevent the fall of all solutions in the population into a local optimum of the solution space. The operator chooses two random integers between 1 and the length of the chromosome, and their values are interchanged.

After each sucessive generation, the best chromosome is evaluated based on the following function, which defines the stopping criteria of our genetic algorithm.

If d is the desired number of sensor nodes per cluster and x is the actual number of sensor nodes per cluster; then the stopping criterion is

$$f(x) = \begin{cases} halt, & x > d-v \text{ and } x < d+v \\ continue, & otherwise \end{cases} \quad (12)$$

Here v , the variance factor, is a configurable number which is kept small so that the deviation of nodes per cluster from the idealised number d is very small.

3   Experiments

The procedure is evaluated using a simulation to solve clustering problems with 200 sensor nodes formed into 5 or 7 clusters. Both distance metrics are evaluated

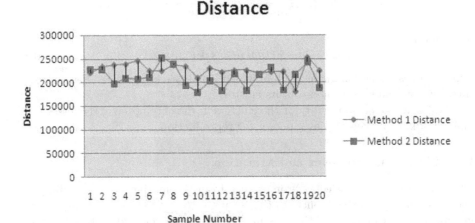

Fig. 1. 200 Sensor Nodes with 5 Clusters: Euclidean Distance vs. Sample Number

Table 1. Results for 200 Sensor Nodes with 5 Clusters. 95% confidence intervals are shown.

Sample Number	Method 1			Method 2		
	Distance	Hops	Generations	Distance	Hops	Generations
1	220848.45	1465	8	227611.97	1766	8
2	234038.78	1544	4	228297.34	1516	5
3	237381.76	1448	8	197320.25	1619	29
4	238598.45	1548	5	209451.14	1485	6
5	245547.93	1570	4	207444.62	1491	11
6	224515.44	1836	6	210619.71	1532	26
7	224621.22	1533	3	252480.4	1750	11
8	236970.6	1481	6	239440.62	1507	4
9	233816.42	1602	7	194193.95	1319	13
10	209097.67	1564	25	178823.69	1210	10
11	229703.67	1500	20	202979.21	1490	9
12	221292.63	1522	12	182316.22	1387	21
13	224854.4	1643	3	217546.27	1446	3
14	224826.93	1204	20	183392.49	1639	6
15	217309.77	1530	6	216654.17	1481	6
16	222321.45	1894	7	232238.75	1322	17
17	222308.27	2067	7	183514.51	1482	13
18	180685.44	1631	24	216261.82	1402	11
19	251970.98	1524	4	242504.42	1532	7
20	224347.01	1445	9	188530.34	1350	6
Total	4525057.27	31551	188	4211621.89	29726	222
Mean	226252.86	1577.55	9.4	210581.09	1486.3	11.1
Standard Dev.	14695.85	181.88	7.01	21857.12	138.17	7.15
CI - alpha 0.05	6440.62	79.71	3.07	9579.13	60.55	3.13

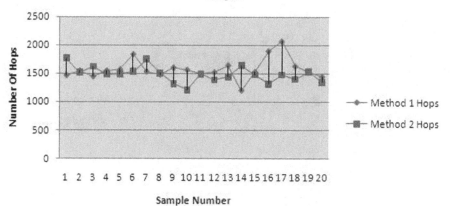

Fig. 2. 200 Sensor Nodes with 5 Clusters: Number of Hops vs. Sample Number

Table 2. Results for 200 Sensor Nodes with 7 Clusters. 95% confidence intervals are shown.

Sample Number	Method 1			Method 2		
	Distance	Hops	Generations	Distance	Hops	Generations
1	207538.79	1235	46	198846.19	1234	62
2	205315.24	1167	48	185865.53	1208	33
3	215053.09	1238	59	211324.77	1216	15
4	223757.73	1238	12	214032.97	1149	56
5	196749.42	1201	46	190492.36	1109	78
6	198045.04	1228	250	209060.44	1208	30
7	210988.72	1163	114	202447.03	1186	70
8	184312.50	1190	154	188157.95	1318	50
9	247205.49	1181	50	246287.19	1092	25
10	206198.46	1109	154	213929.99	1309	167
11	225504.23	1125	74	239173.94	1203	94
12	215867.22	1059	153	211505.73	1097	16
13	189255.92	1162	143	192837.43	1200	38
14	212369.66	1299	110	188978.95	1165	71
15	217140.29	1317	44	201937.53	1222	118
16	224696.02	1242	206	200896.1	1334	35
17	227468.38	1240	89	211490.07	1130	52
18	209863.67	1111	129	207901.23	1108	146
19	225355.39	1146	62	225005.72	1118	80
20	211554.06	1371	116	201065.67	1144	200
Total	4254239.32	24022	2059	4141236.79	23750	1436
Mean	212711.97	1201.10	102.95	207061.84	1187.50	71.80
Standard Dev.	14500.10	76.07	61.25	15975.85	72.72	50.90
CI - alpha 0.05	6354.83	33.34	26.84	7001.60	31.87	22.31

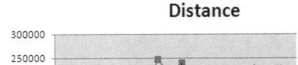

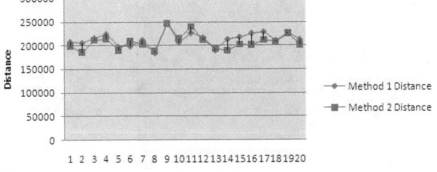

Fig. 3. 200 Sensor Nodes with 7 Clusters: Euclidean Distance vs. Sample Number

Hops

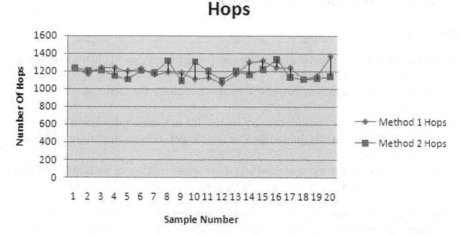

Fig. 4. 200 Sensor Nodes with 7 Clusters: Number of Hops vs. Sample Number

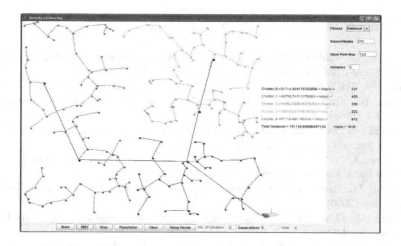

Fig. 5. 200 Sensor Nodes with 5 Clusters Simulation Results

using 20 randomly generated samples. Table 1 shows the results of 200 sensor nodes formed into 5 clusters with a variance factor of 5, 80% crossover probability and population size of 100. Similarly, Table 2 shows the results of 200 sensor nodes formed into 7 clusters with a variance factor of 4, 80% crossover probability and population size of 100. In both the tables Method 1 refers to fitness function 1 and Method 2 refers to fitness function 2. Both methods are evaluated using total distance and number of hops as performance measures.

The simulation takes a randomly generated population as input and in turn computes the fitness function for each chromosome. The best chromosome (elitism) is displayed after every generation by the visualization tool as illustrated in Figure 5.

4   Conclusion and Future Work

This work

1. Establishes that a genetic algorithm can produce very good clusters of nodes in Wireless Sensor Networks.
2. The Voronoi diagram is an effective methodology for defining the cluster boundaries.
3. The methodology accomodates any metric for measuring distances from sensor nodes to relay heads, including Euclidean distances or minimum spanning tree distances.
4. Our experimental work was conducted for sensor networks with 300 and 500 sensor nodes, and consistent and high-performance results were obtained.

There are many avenues for future work. One important future improvement is to take into account the node failure and the relay node death and also to include the transmission ranges of each node.

References

1. Tasoulis, D., Vrahatis, M.: Novel approaches in unsupervised clustering. In: NEMIS 2004 Final Conference in Knowledge Mining, Patras, Greece (2004)
2. Agarwal, P.K., Procopiuc, C.M.: Exact and approximation algorithms for clustering. In: 9th Sysposium on Discrete Algorithms, San Francisco, California, pp. 658–667 (1998)
3. Megiddo, N., Supowit, K.: On the complexity of some common geometric problems. SIAM Journal on Computing 13, 182–196 (1984)
4. Werner, J.C., Fogarty, T.C.: Genetic algorithm applied in clustering datasets. South Bank University, London (2002)
5. Painho, M., Bacao, F.: Using genetic algorithms in clustering problems. GeoComputation (2000)
6. Rourke, J.: Computational Geometry in C, 2nd edn. Cambridge University Press, Cambridge (1998)
7. Vieira, M.A., Vieira, L.F., Ruiz, L.B., Loureiro, A.A., Fernandes, A.O., Nogueir, N.: Scheduling nodes in wireless sensor networks: A Voronoi approach. In: 28th Annual IEEE International Conference on Local Computer Networks (LCN 2003) (2003)
8. Preparata, F.P., Shamos, M.I.: Computational Geometry An Introduction. Springer, New york (1985)
9. Minimum Spanning Tree (1997),
   http://www2.toki.or.id/book/AlgDesignManual/BOOK/BOOK4/NODE161.HTM
10. Fitness Proportionate Selection (2007),
    http://en.wikipedia.org/wiki/Fitness_proportionate_selection

VGTR: A Collaborative, Energy and Information Aware Routing Algorithm for Wireless Sensor Networks through the Use of Game Theory

Alexandros Schillings and Kun Yang

School of Computer Science and Electronic Engineering, University of Essex, UK
{aschil,kunyang}@essex.ac.uk

Abstract. Game Theory (GT) is a branch of applied mathematics that models situations where players (participants in a game) participate in a strategic situation (the game) in which they perform different actions attempting to maximise their profits, while at the same time minimise losses. As nodes in Wireless Sensor Networks (WSN) can be abstracted as the players in such games where energy and information are valuable resources it is obvious that Game Theory provides a solid framework for both the modelling and the induction of node behaviour in such networks. The proposed algorithm induces an energy-aware and efficient collaborative behaviour to the nodes using sensor centric information, by making them aware of their interdependency, without compromising the main purpose of the network - the collection of information.

1 Introduction

A Wireless Sensor Network (WSN) is composed of a collection of wireless nodes that are designed to monitor, store and report phenomena, usually with minimal human interaction [1]. They are usually deployed as part of a set-and-forget strategy where an operator is only involved to collect data from a designated node called Sink.

Data collection is the main purpose for utilising a WSN. Nodes are built with a specialised sensing task in mind; they are then deployed and are expected to convey their findings to a data collection point, i.e., the Sink. This paper adopts a query-driven methodology to accomplish the extraction of data from the network [2]. To accomplish this, a query is broadcast from the Sink towards the required nodes. Then all the nodes that receive the query respond by sending their sensed data back to the Sink, resulting in multiple reporting sensors (sources) and one receiving node (Sink). This can be naturally expressed by a tree whose root is the Sink, the leaves are the multiple sources, and the intermediate nodes are data relaying nodes. We call this tree *Data Routing Tree* (DRT). Each branch in the DRT constitutes a route starting from a reporting sensor and ending at the Sink. The problem is how to construct such a DRT so as to preserve nodes' energy to prolong the lifetime of the entire WSN.

While the majority of the existing work [2-6] adopt a conventional approach to solve the data aggregation and routing problems in WSNs, this paper proposes an interdisciplinary approach where game theory is utilized to tackle these challenges.

N. Trigoni, A. Markham, and S. Nawaz (Eds.): GSN 2009, LNCS 5659, pp. 51–62, 2009.

Game theory (GT) is a framework that allows the modelling of multiparty decision problems and is increasingly attracting more attention as a mechanism to solve various problems in wireless networks [7-10]. R. Kannan *et al* [7] propose an algorithm that induces the formation of a maximally reliable data aggregation tree, in both geographically aware and agnostic networks. In the proposed approach, the routing decisions occur on a hop by hop distributed basis. A. Urpi *et al* [8] developed a GT model that describes the collaboration of selfish wireless nodes which can be applied in multiple scenarios. Márk Fèlegyhazi *et al* [9] propose a theoretical model in which nodes do not have an incentive to cooperate but their behaviour results in a Nash Equilibrium (NE).

We propose a GT based approach to induce the creation of a DRT in which the nodes will cooperate to route information. In this tree nodes are the players and survivability is the resource over which the nodes compete. Nodes are aware that their actions and choices affect other nodes that are located upstream. Upstream nodes of a node are these nodes that link the node to the Sink. This creates a drive to preserve these nodes as without them a downstream node will be unable to perform its task. We expand the notion of network survivability [5], and consider it a result of node survivability. Nodes can improve their survivability by looking after upstream nodes in as to avoid segmentation. We call the resulting protocol *Versatile Game Theoretic Routing Protocol (VGTR)*.

The rest of this paper is organised as follows: Section 2 introduces the game theoretic concepts that are used for this algorithm and formulates the problem while Section 3 describes the proposed solution in game theoretic terms and introduces the proposed algorithm. Section 4 discusses the evaluation results before the concluding remarks in Section 5.

2   Preliminaries

2.1   Network Assumptions and Design Choices

During the design of VGTR, the design goals were considered: energy efficiency, scalability, versatility, information awareness and practicality.

In order to facilitate these goals, the following design choices are made:

1. The protocol is query based. This eliminates any traffic unless data is requested.
2. Data is routed along a single, dynamically established, path. When a node needs to send data, it selects a neighbour to send the message to.
3. The routing protocol is built around the concept of minimising network segmentation in order to prolong the network's effective lifetime. This is achieved by abstracting the network in a way that each node is aware of issues between itself and the sink with no knowledge of the actual topology.

Similarly, the following assumptions were made regarding the network capabilities:

1. The nodes have the same maximum battery capacity, even though their hardware design can be heterogeneous.
2. The network nodes are assumed to have identical transmitters and receivers. This is to ensure that if node n_i can contact node n_j, then the opposite will be true.

3. The sink is assumed to have an infinite power supply.
4. The transmission range is fixed and is small enough so that most nodes will be unable to reach the sink without hopping at least once.
5. Nodes have a general idea of their position in the network and the position of the sink. This can be achieved as simply as using a hop count metric or by more sophisticated means.
6. Nodes cannot drop packets they are asked to forward.

2.2 Problem Re-investigation

Nodes are the basic building blocks that a WSN is composed of and each plays the dual role of both sensing and conveying information to the Sink from other nodes. In order to fulfil these tasks, they need to stay alive for as long as possible by conserving energy. On the other hand, in order to convey the information, they need to expend energy

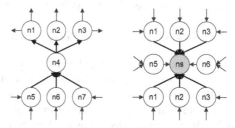

Fig. 1. Bridging Node   **Fig. 2.** Sink Hotspot

and act as liaisons between other nodes and the Sink, thus jeopardising their survival. This conflict of interest between the two main drives of a node (survival and purpose) makes GT suitable for WSNs. Nodes normally make decisions only concerning their next hop. This causes problems since upstream nodes that will be contacted during the data upload are potentially more important to the network due to their position and place in the topology, as illustrated by node n_4 in Fig. 1. Another type of bridging nodes is those around the Sink, e.g., n_1-n_8 as illustrated in Fig. 2. When making next-hop selection, nodes try to switch away from areas that are becoming "hot" in favour of more stable ones. Due to their importance, these nodes are described as critical.

Most current routing approaches use a combination of path cost metrics and hop count to find the best path for the nodes to use [1]-[5]. These methods can be inefficient in case of multiple queries or increased data load as they do not confer enough information concerning the state of the network. An ideal approach would allow the nodes to make decisions based on the results of their past strategies.

3 Game Theoretic Modelling of Collaborative and Energy Efficient Data Routing

The data routing problem described in Section 1 can be formally formulated as follows: a WSN of n responding sensor nodes is represented by a normal form game $G = \{S_1, ..., S_n; u_1, ..., u_n\}$ where the strategy space that node i can select from is represented by $S_i(i = 1, ..., n)$ and where the corresponding payoff function of node i on strategy space S_i is represented by $u_i(i = 1, ..., n)$. We need to formulate the payoff functions in a way that will help node i select a strategy S_i that represents the best response to the strategies selected by the other n-1 nodes. The resulting strategy (also called a strategy profile) $s = \{s_1, \ldots, s_n | s_i \in S_i, i = 1, \ldots, n\}$ needs to place the nodes responding to the query in a Nash Equilibrium (NE).

The nature of the payoff functions and the subsequent selection of strategies will lead to the formation of s which is carried out during the creation of a DRT T rooted at the node originating the query (the Sink n_s). The tree T is consisted of nodes that respond to the query and as such is a subgraph of a graph that contains all nodes in the network and the possible edges for those nodes. Each edge l_{ij} connects two nodes n_i and n_j only if node n_i and n_j are within each other's transmission range. The formation of T should happen in a way which intelligently judges the energy consumption of the paths and takes notice of nodes with low remaining energy or high information value.

3.1 Network Abstraction and Energy Awareness

In order to facilitate a protocol with the parameters described above, three payoff functions are used. The first two are representative of node survivability, while the third one represents the importance of the information collected. Node survivability represents the capability of a node to remain in contact with the Sink for as long as possible. It is directly affected by the health of the upstream network and, to a greater extent, the health of a node's neighbours. The lower the energy dissipation rate of these factors the higher the survivability of a given node. By definition, a node wants to remain connected for as long as possible. To achieve this, it needs to protect its next hop neighbours and all the possible paths behind them that lead to the sink. This common interest between the nodes is what makes collaboration an efficient strategy. Thus, the payoff is represented by two factors: *UPPN Robustness* and *Neighbour Robustness*.

$UPPN_{n_i n_j}$ *(Upstream Potential Path Nodes)* is defined as the set of nodes that exist between node n_i and the sink assuming that n_i will use upstream node n_j as it's next hop. Each node has a UPPN set for each of its upstream neighbours and it is possible the UPPN sets to intersect.

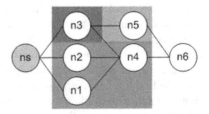

Fig. 3. Upstream Potential Path Nodes (UPPN) Example

For example in Fig. 3, node n_6 has two UPPN sets: $A_{n6n4} = \{n_1, n_2, n_3, n_4\}$ and $A_{n6n5} = \{n_3, n_5\}$ where $A_{n6n4} \cap A_{n6n5} = \{n_3\}$. Node n_i is rarely aware of the exact nodes in each of its UPPN sets but it is not required to be either. The UPPNs are formed during the query phase, as the query propagates itself in the network.

In this paper, energy is expressed differently from traditional depiction. Instead of expressing it with an absolute representation (using Joules for example), a time derivative, representing the amount of time a node has left based on past workload, is used. This is the equivalent of asking a car driver how much fuel is left. Most drivers would reply using a unit of distance ("*about 50 km*") instead of volume ("*10 litres*") as it is more informative since volume alone does not convey any practical meaning in this context, as the "value" of a litre is dependent on additional factors.

In the following text, whenever the phrase "the nodes behind node n_j" is met, it should be read as "the nodes belonging to $UPPNn_in_j$".

UPPN Robustness. One of the factors that directly affects the lifetime of a network is the health of the paths hidden behind its next hop neighbours. By selecting its next hop, a node not only affects the hop itself, but sets in motion a chain of transmissions to which it is oblivious. Path criticality is affected by the changes in average energy of all the nodes in the path and especially the changes in the average energy of the nodes reported as having the least energy in them.

In order to calculate this, the lifetime of the path needs to be approximated. It is assumed that node n_j is connected to the sink via x neighbours (thus x UPPNs) and that $U = UPPN_1 \cup UPPN_2 \cup ... UPPN_{x-1} \cup UPPN_x$.

1. The Average Energy of all nodes in U (AEU). This is gathered by averaging the energy level of all nodes the query passes through, or more simply by averaging the energy of the nodes belonging to the UPPNs of a node (1):

$$AEU^{nj} = \frac{1}{|U|}\sum_{y=1}^{|U|} E^{n_y}, \ if \ n_y \in U \tag{1}$$

The Energy of the node with the absolute minimum energy in all the UPPNs n_j is connected to (MEU). This is essentially, the single lowest value found in each of the received query copies (2):

$$MEU^{nj} = Min(E_{n1}, E_{n2}, \cdots, E_{n|U|-1}, E_{n|U|}), \ \forall n \in U \tag{2}$$

The Average Energy of nodes reported as having Minimum energy (AME). As each node can receive multiple copies of a query from its neighbours, by averaging MEU forwarded by each query copy we can produce AME. Thus, assuming that node n_j is connected to the sink via x neighbours and A is the set of valid upstream neighbours, then the value reported to n_i will be the following (3):

$$AME^{nj} = \frac{1}{x}\sum_{y=1}^{x} MEU^{n_y}, \ if \ n_y \in A \tag{3}$$

Collectively AEU, MEU and AME are part of the Abstracted Network Status Attributes (ANSAs) which are used to provide nodes with an overview of the network status by abstracting the network state as to minimise traffic and overhead. Later in this paper, we will introduce additional ANSAs.

By calculating the rate of change for each of the above values, it is possible to approximate the expected lifetime of each of these node groups in terms of rounds based on the current energy characteristics of the path ((4), (5) and (6)).

$$D^{AEU} = E^{AEU} / \frac{d_{AEU}}{dt} \tag{4}$$

$$D^{AME} = E^{AME} / \frac{d_{AME}}{dt} \tag{5}$$

$$D^{MEU} = E^{MEU} / \frac{d_{MEU}}{dt} \tag{6}$$

In order to produce the path's overview, it is important to note that the path might *possibly* function only for the D^{MEU} duration before network segmentation occurs, as the node producing the D^{MEU} value is possibly a critical node. This is represented in formula (7).

$$W_{ij}^{E(UPPN)} = D_{ij}^{MEU} \frac{D_{ij}^{AME}}{D_{ij}^{AEU}} \tag{7}$$

As the nodes forming D_{ij}^{AME} are a node subgroup of those for D^{AEU}, their ratio is of the uniformity of the energy dissipation in the path. The healthier a path is, the er W_{ij}^{Path} will be to D_{ij}^{AEU}, thus signifying a uniform energy discharge. D_{ij}^{MEU} is used as the key value since if the node is located at a critical spot, then all the other metrics are meaningless.

In order to fully integrate the W_{ij}^{UPPN} value to the payoff, it must be normalised to the more practical scale [0, 100] (8):

$$F_{ij}^{E(UPPN)} = 100 * \frac{W_{ij}^{UPPN} - D^{MIN}}{D^{MAX} - D^{MIN}} \tag{8}$$

where D^{MAX} and D^{MIN} are the maximum and minimum possible node lifetime in rounds respectively.

Neighbour robustness. Nodes can only directly affect the energy level of their next hop neighbours and it is unable to measure the load that is imposed on its next hop by other nodes. Thus it can only estimate the actual energy cost the next hop node pays.

Until enough data has been collected to start deriving, the node will try to spread the packets it needs to forward between its neighbours. After z rounds have passed, it will then try to keep the rate of the energy expenditure of each individual neighbour (D^{NE}) as low as possible by trying to uniform the expenditure by comparing it with projected lifetime of the higher energy neighbour (D^{MNE}). In order to do this, each node attempts to maintain the ratio given in (9) as close to 1 as possible:

$$W_{ij}^A = \frac{D_j^{NE}}{D_i^{MNE}} \tag{9}$$

where:

$$D^{NE} = E^{NE} / \frac{\delta_{NE}}{\delta t} \tag{10}$$

$$D^{ANE} = E^{MNE} / \frac{\delta_{MNE}}{\delta t} \qquad (11)$$

Again, the W_{ij}^A value will need to be normalized in a [0,100] range (12):

$$F_{ij}^A = 100 * W_{ij}^A \qquad (12)$$

3.2 Information Value and Data Aggregation

As mentioned before, the main purpose of a WSN is the collection and reporting of information. Since we are using a query-based protocol, the collection of data can be asynchronous of the request. Each node records information of some sort, depending on the application of the WSN. Assuming that a node can calculate and assign a value corresponding to the importance (or quality) of the information stored on it when it receives a query, then it is possible for nodes to convey to downstream nodes the Information Value (IV) of the current path in order to provide more input for the formation of the DRT. The mechanics behind the calculation method is application specific but since IV is application agnostic, it can support different types of data in the same network. For VGTR, we assume that the IV assigned to data can have a range of [0, 100] with 100 being the most important.

The calculation of IV of a given path is similar to the calculation of the Node Survivability and thus additional ANSAs will be used. Unlike Node Survivability, we are concerned with the high end of the measurement spectrum (higher values instead of low). It is assumed that node n_j is connected to the sink via x neighbours (thus x UPPNs) and that $U = UPPN_1 \cup UPPN_2 \cup ... UPPN_{x-1} \cup UPPN_x$.

1. The Average IV of all nodes in the UPPN (AIU). This is gathered by averaging the IV of all nodes the query passes through, or more simply by averaging the IV of the nodes belonging to the UPPNs of a node. (13) :

$$AIU^{nj} = \frac{1}{|U|}\sum_{y=1}^{|U|} IV^{n_y}, \quad if\ n_y \in U \qquad (13)$$

2. The IV of the node with the absolute *maximum* IV in all the UPPNs n_j is connected to (MIU). This is the single lowest value found in the received query copies. (14):

$$MIU^{nj} = Max(IV_{n1}, IV_{n2}, \cdots, IV_{n|U|-1}, IV_{n|U|}), \quad \forall n \in U \qquad (14)$$

3. The Average IV of nodes reported as having Maximum IV (AMI). As each node can receive multiple copies of a query from their neighbours, by averaging MIU forwarded by each query copy we can produce AMI. Thus, assuming that node n_j is connected to the sink via x neighbours and A is the set of valid upstream neighbours, then the value reported to n_i will be the following (3):

$$AMI^{nj} = \frac{1}{x}\sum_{y=1}^{x} MIU^{n_y}, \quad if\ n_y \in A \qquad (15)$$

Contrary to formulas (4), (5) and (6), we consider the most prosperous path the one that has its *MIU* higher than others and the F_{ij}^{IV} ratio (16) closest to 1.

$$F_{ij}^{IV(UPPN)} = MIU^{nj} \frac{AIU^{nj}}{AMI^{nj}} \tag{16}$$

As IV cannot be extrapolated from past results, a time derivative is not used when calculating F_{ij}^{IV} and as it is already in the [0,100] range, no normalisation is necessary.

3.3 Overall Node Payoff

The following formula provides the overall payoff for node n_i to choose node j for the next hop and can be calculated as (17):

$$F_{ij}^{Tot} = \begin{cases} \alpha * F_{ij}^{E(UPPN)} + \beta * F_{ij}^{A} + \gamma * F_{ij}^{IV(UPPN)}, & \text{if } s_j \in T \\ 0, & \text{otherwise} \end{cases} \tag{17}$$

Where α, β, γ are weights given to each payoff function ($\alpha+\beta+\gamma=1$ and $\alpha\geq0$, $\beta\geq0$, $\gamma\geq0$). By altering the weightings, we can cause a node to take into account only short term considerations ($\beta=1$), be completely altruistic ($\alpha=1$), seek only the highest IV with no regards to energy ($\gamma=1$) or any possible strategies in between.

3.4 Algorithm Operation Overview

Step 1 (Query Broadcast): The Sink (ns) initiates the process by transmitting a query towards the nodes that it is interested in. The query packet contains (along with the actual query) any weighting the Sink will want to provide in order to bias the decisions of the nodes, an optional TTL value and the values needed for the payoff functions (Fig. 4.).

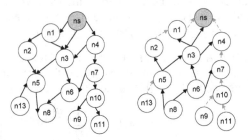

Fig. 4. Flooding Process   **Fig. 5.** Possible Decisions

Step 2 (Query Propagation): Each node receiving the query waits a predetermined amount of time (in order to receive copies of the same query), updates the ANSA fields accordingly and forwards the query to its downstream neighbours.

Step 3 (Query Reply): Once the target area is reached (or *TTL=0*) the nodes that received the query last are able to calculate the possible outcomes of choosing each of the possible upstream branches and they can estimate the values of the average and worst case upstream node. This will allow each node to calculate its strategy space S_i assign a probability p_{ik} to each strategy and make a selection (Fig. 5.), based on the query's $\alpha\beta\gamma$ values. This procedure is repeated on each upstream node that will actually participate (thus forming T). If a path seems to deplete its energy faster than others, then the probability p_{ik} that leads to its selection is reduced accordingly.

4 Algorithm Evaluation

4.1 Experiment Setup

The following protocols have been chosen as benchmarks:

1. EAR [13], as it can be considered as a simple, non GT progenitor of VGTR.
2. SEER [14] as it is a newer, simple, event-driven multihop protocol that is starting to appear in literature.
3. Directed Diffusion (DD) [15], as it a popular data-oriented protocol, which would be useful to benchmark the IV capabilities of VGTR.
4. LEACH [6], as it is used as a benchmark in various publications and, due to its highly constrained but efficient nature, it represents a high-end benchmark.

Table 1. VGTR Settings

	α	β	γ
VGTR1	0.5	0.5	0
VGTR2	0.5	0.25	0.25
VGTR6	0	0	1

For the simulations, the ns-2 network simulator v2.30 [16] is used. Each node n_i is assigned a random initial energy and a random IV, while all nodes are stationary. For the time derivative, a history of 5 samples is taken, while the transmission range for all the nodes and the sink is 15. The topology used was a square 100x100 "arena" which was divided into 5x5 cells in which 100 sensor nodes were uniformly distributed. The three main VGTR variants used the settings shown on Table 1.

4.2 Experimental Results

Both DD and SEER deplete their energy supplies quite early in the simulations. This occurs because DD exhibits limited energy awareness when compared with its information retrieval capabilities. SEER, even though it is energy aware, its reliance on hop count causes it to easily generate critical nodes and hot paths. This is partially off-set by the lack of query broadcast packets in the network. EAR and VGTR fare better, as they attempt to balance the load between multiple paths (if available). VGTR nodes, in addition, are able to detect hotpaths and critical nodes, as they are formed, and will minimise their use if other routes are available (Fig. 6., Table 2.).

Although VGTR does not allow accurate targeting, it does provide a concept of direction and depth, and as such the retrieval of information from areas (as opposed to nodes) is possible. SEER and LEACH being source initiated protocols (LEACH can be loosely described as such) do not offer direct targeting.

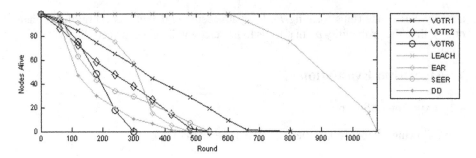

Fig. 6. Protocol Lifetime Comparison

Table 2. Protocol Performance

	VGTR1	VGTR2	VGTR6	DD	SEER	LEACH	EAR
First node death	19	51	58	42	61	619	46
Last node death	622	481	261	421	492	1074	411
Rate of node death (total)	0.16	0.21	0.38	0.24	0.20	0.09	0.24
Rate of node death (after first death)	0.16	0.23	0.49	0.26	0.23	0.22	0.27
Information Value	~49%	~72%	~75%z	~68%	~48%	~49%	~49%

When comparing the extraction capabilities, we split the network into three areas:

1. The sink neighbourhood, which contains all nodes directly connected with the sink.
2. The target neighbourhood, which contains the target nodes' neighbours.
3. The path nodes, which include all nodes that are between the target and the sink.

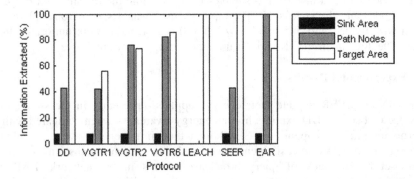

Fig. 7. Information Extraction Comparison

Directed Diffusion offers the best results as it is a protocol designed for the efficient extraction of information. It is then followed by VGTR and SEER with LEACH being last. The main difference between VGTR and SEER can be located to the fact that SEER will always report maximum information from the target area but will fair average in the path nodes segment as it uses no metrics to evaluate the information stored in the intervening nodes. VGTR on the other side will attempt to maximise the

IV gained at the target area, and also aim to maximise the information that will be extracted from the intervening nodes. Since multiple simultaneous active paths will be formed during a query, the resulting IV level is higher (Fig. 7, Table 2.).

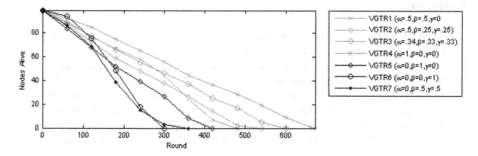

Fig. 8. VGTR Variant Comparison

Fig. 8. and Table 3 contrast the performance of different VGTR variants in both their energy-aware and their information retrieval aspects.

Table 3. Node Death Analysis

	VGTR1	VGTR2	VGTR3	VGTR4	VGTR5	VGTR6	VGTR7
First node death	19	51	11	9	5	22	7
Last node death	622	481	531	443	410	261	351
Rate of node death (total)	0.16	0.21	0.19	0.22	0.24	0.38	0.28
Rate of node death (after first death)	0.16	0.23	0.19	0.23	0.24	0.38	0.29
Information Value	~49%	~72%	~72%	~48%	~49%	~75%	~50%

Depending on the α, β and γ values diverse performance can be achieved in order to accomplish the task at hand, generally by trading off life expectancy for improved information value. Naturally, these choices can be made as needed during the query phase of the protocol.

5 Conclusions

In this paper we introduce an energy aware Game Theoretic algorithm that induces and maintains collaboration in WSNs. This is done by making nodes aware of the results of their actions and forcing them to rotate the selection of their next hop in a calculated way by utilising the payoff functions in order to select one that will extend their functional life time, as well as selecting prosperous paths (information-wise) while taking care of bridging nodes. One thing that still warrants investigation is the automatic, on the fly, calculation of the benefit weightings (α, β, γ) based on the current energy of the network and past moves. This will allow the nodes to adapt even faster to situations and avoid making moves that are mathematically correct but strategically unsound.

Acknowledgement. The work presented in this paper was funded by the UK EPSRC CASE studentship.

References

1. Akyildiz, I., Su, W., Sankarasubramaniam, Y.: A Survey on Sensor Networks. IEEE Communication Magazine, 102–114 (August 2002)
2. Tilak, S., Abu-Ghazaleh, N.B., Heinzelman, W.: Taxonomy of Wireless Micro-Sensor Network Models. ACM Mobile Computing and Communication Review 6(2) (2002)
3. Jiang, Q., Manivannan, D.: Routing Protocols for Sensor Networks, Department of Computer Science, University of Kentucky, Lexington, KY 40506. IEEE, Los Alamitos (2004)
4. Al-Karaki, J.N., Kamal, A.E.: Routing Techniques in Wireless Sensor Networks: A Survey" Dept. of Electrical and Computer Engineering, Iowa State University, Ames, Iowa 50011
5. Shah, R.C., Rabaey, J.M.: Energy Aware Routing for Low Energy Ad Hoc Sensor Networks. In: Proc. IEEE Wireless Communication and Network Conference, March 2002, vol. 1, pp. 350–355 (2002)
6. Heinzelman, W.R., Chandrakasan, A., Balakrishnan, H.: Energy Efficient Communication Protocol for Wireless Microsensor Networks. In: Proceedings of the Hawaii International Conference on System Sciences, Maui, Hawaii (January 2000)
7. Kannan, R., Sitharama Iyengar, S.: Game Theoretic Models for Reliable Path-Length and Energy Constrained Routing with Data Aggregation in Wireless Sensor Networks. IEEE Journal on Selected Areas in Communication 22 (August 2004)
8. Urpi, A., Bonuccelli, M., Giordano, S.: Modelling cooperation in mobile ad hoc networks: a formal description of selfishness. In: Proceedings of WiOpt Workshop (2003)
9. Fèlegyhazi, M., Hubaux, J.-P., Buttyán, L.: Nash Equilibria of Packet Forwarding Strategies in Wireless Ad Hoc Networks. IEEE Transactions on Mobile Computing 5(5) (May 2006)
10. Wang, W., Li, X.-Y.: Low-Cost Routing in Selfish and Rational Wireless Ad Hoc Networks 5(5) (May 2006)
11. Game Theory for Applied Economists, Game Theory for Applied Economists. Princeton University Press, Princeton New Jersey, ISBN0-691-00395-5
12. Sha, K., Shi, W.: Modelling the Lifetime of Wireless Sensor Networks. Sensor Letters 3, 1–10 (2005)
13. Yuen, W.H., Sung, C.W.: On Energy Efficiency and Network Connectivity of Mobile Ad Hoc Networks. In: Proceedings of the 23rd International Conference on Distributed Computing Systems (ICDCS 2003), Providence, Rhode Island, USA, May 19-22 (2003)
14. Hancke, G.P., Leuschner, C.J.: SEER: A Simple Energy Efficient Routing Protocol for Wireless Sensor Networks. SACJ 39, 17–24 (2007)
15. Intanagonwiwat, C., Govindan, R., Estrin, D.: Directed diffusion: A scalable and robust communication paradigm for sensor networks, August 2000, pp. 56–67 (2000)
16. The NS-2 Network Simulator, http://www.isi.edu/nsnam/ns/

Building Efficient Aggregation Trees for Sensor Network Event-Monitoring Queries

Antonios Deligiannakis[1], Yannis Kotidis[2], Vassilis Stoumpos[3], and Alex Delis[3]

[1] Technical University of Crete
[2] Athens University of Economics and Business
[3] University of Athens

Abstract. In this paper we present algorithms for building and maintaining efficient aggregation trees that provide the conduit to disseminate data required for processing monitoring queries in a wireless sensor network. While prior techniques base their operation on the assumption that the sensor nodes that collect data relevant to a specified query need to include their measurements in the query result at every query epoch, in many event monitoring applications such an assumption is not valid. We introduce and formalize the notion of event monitoring queries and demonstrate that they can capture a large class of monitoring applications. We then show techniques which, using a small set of intuitive statistics, can compute aggregation trees that minimize important resources such as the number of messages exchanged among the nodes or the overall energy consumption. Our experiments demonstrate that our techniques can organize the data aggregation process while utilizing significantly lower resources than prior approaches.

1 Introduction

Many pervasive applications rely on sensory devices that are able to observe their environment and perform simple computational tasks. Driven by constant advances in microelectronics and the economy of scale it is becoming increasingly clear that our future will incorporate a plethora of such sensing devices that will participate and help us in our daily activities. Even though each sensor node will be rather limited in terms of storage, processing and communication capabilities, they will be able to accomplish complex tasks through intelligent collaboration.

Nevertheless, building a viable sensory infrastructure cannot be achieved through mass production and deployment of such devices without addressing first the technical challenges of managing such networks. In this paper we focus on developing the necessary data aggregation infrastructure for supporting aggregate queries. For such applications, most recent proposals rely on building some type of ad-hoc interconnect for answering a query such as the *aggregation tree* [16,26]. This is a paradigm of in-network processing that can be applied to non-aggregate queries as well [7]. In this paper we concentrate on building and maintaining efficient *aggregation trees* that will provide the conduit to disseminate all data required for processing aggregate queries, while minimizing important resources such as the number of messages exchanged among the nodes or the overall energy consumption.

N. Trigoni, A. Markham, and S. Nawaz (Eds.): GSN 2009, LNCS 5659, pp. 63–76, 2009.

While prior work [4,23,24] has also tackled similar problems, previous techniques base their operation on the assumption that the sensor nodes that collect data relevant to the specified query need to include their measurements (and, thus, perform transmissions) in the query result at every query *epoch*. However, in many monitoring applications such an assumption is not valid. Monitoring nodes are often interested in obtaining aggregate values only from sensor nodes that detect interesting events. In such applications, each sensor node is not forced to include its measurements in the aggregate at each epoch, but rather such a *query participation* is evaluated on a per epoch basis, depending on its readings and the definition of interesting events. In this paper we term the monitoring queries where the participation of a node is based on the detection of an event of interest as *event monitoring queries* (EMQs).

Our techniques base their operation on collecting simple statistics during the operation of the sensor nodes. The collected statistics involve the number of events (or, equivalently, their frequency) that each sensor detected in the recent past. Our algorithms utilize these statistics as hints for the behavior of each sensor in the near future and periodically reorganize the aggregation tree in order to minimize certain metrics of interest, such as the overall number of transmissions or the overall energy consumption in the network. The formation of the aggregation tree is based on the aggregation and local transmission of only a small set of values at each node termed as *cost factors* in our framework. Using these cost factors each sensor selects its parent node, through which it will forward its results towards the base station, based on the estimated corresponding *attachment cost*. In a nutshell, the attachment cost of a parent selection is the increase in the objective function (i.e., the number of transmitted messages) resulting from this selection. Given the estimates of attachment costs that our algorithms compute, our work demonstrates that they are able to design significantly better aggregation trees than existing techniques.

Our contributions are summarized as follows:

1. We formally introduce the notion of EMQs in sensor networks. EMQs are a superset of existing monitoring queries, but are handled uniformly in our framework, irrespectively of the minimization metric of interest.
2. We present detailed algorithms for minimizing important metrics such as the number of messages exchanged or the energy consumption during the execution of an aggregate EMQ. The presented algorithms are based on the aggregation and transmission of a small, and of constant size, set of statistics. We introduce our algorithms along with a succinct mathematical justification.
3. We present a detailed experimental evaluation of our algorithms. Our results demonstrate that our techniques can achieve a significant reduction in the number of transmitted messages, or the overall energy consumption, compared to alternative algorithms.

2   Related Work

The database community has long been the advocate of using an embedded database management system for data acquisition in sensor networks [16,26]. The use of a declarative SQL-like query interface allows rapid development of applications in such systems without the need to manage hand-coded programs at each sensor node [17].

In the database community different types of popular queries have been discussed, such as aggregate [5,6,16,21,19], join [2], model-based [9,14] and select-all queries [7,22]. Tracking queries that seek to determine the spatial extent of a particular phenomenon have also been considered [10,25]. In [16] the nodes are first organized in a tree topology, termed the aggregation tree. During query execution, each epoch is subdivided into intervals and parent nodes in the aggregation tree listen for messages containing partial aggregates from their children nodes during pre-defined time-slots. Another notable method for synchronizing the transmission periods of nodes is the recently proposed wave scheduling approach of [8]. The work in [28] describes a framework that profiles recent data acquisition activity by the nodes and computes their waking window though an in-network execution of the critical path method. This technique is complementary to our work, as they help identify a proper scheduling for data transmission by the nodes, while our methods focus on optimizing the routing topology.

Many of the low-level networking details have already been discussed in the networking community and, thus, can be utilized in our framework. As an example, nodes in unattended wireless networks must be able to self-configure [3] and discover their surrounding nodes [11]. Prior work on computing energy-efficient data routing paths (such as the aggregation tree) [13,23,24] have tackled similar problems, but these techniques base their operation on the assumption that the sensor nodes that collect data relevant to the specified query need to include their measurements in the query result at every query epoch. However, this assumption does not hold in event monitoring queries that are the scope of our framework. In the other end of the spectrum, the work in [15] and [12] discuss join and aggregation queries involving rare events. Thus, they follow an alternative path, which is to construct the data collection network on-the-fly when such events occur. However, this practice in unsuitable for our setting involving sensor nodes with both low and high participation frequencies, since it would incur a high overhead for frequently maintaining the collection network. Furthermore, the work in [12] assumes the existence of a high speed connection for all nodes at the boundaries of the network, through which the data that reaches the boundary nodes can be communicated.

3   Motivational Example

In Table 1, we present examples of the two main classes of monitoring queries in sensor networks. We borrow the syntax of TinyOS [16] to denote the epoch duration (e) and the lifetime of the query (t). The predicate *inclusionConditions* has been added in order to

Table 1. An Aggregate Query over the Values Collected by Sensor Nodes

Aggregate Query
SELECT   AggrFun(s.value)
FROM   Sensors s
WHERE   inclusionConditions(s) = true
SAMPLE PERIOD e FOR t

specify which sensor nodes will participate in the query evaluation per epoch. At each query epoch, all the sensor nodes that include their collected data in the query result are termed in our framework as *epoch participating nodes*. For queries that wish to collect data from all the sensor nodes at each epoch, the above predicate always evaluates to *true*.

When a monitoring query specifies inclusion predicates, these may contain either static or dynamic predicates (or both) regarding the sensor nodes. Examples of static predicates may involve, but are not limited to, the collection of measurements from: (i) Sensors with specific identifiers; (ii) Immobile sensors in a specific area; or (iii) Sensors monitoring a specific quantity, in cases of sensor networks with diverse types of sensor nodes that monitor different quantities. Static predicates are very useful in a variety of applications and have received the focus of the bulk of past research [16,26]. Inclusion conditions that contain only static predicates result in a fixed subset of the sensor nodes participating in the query output at each epoch. This allows for simple data dissemination and collection protocols based on fixed aggregation trees that need to be altered only when either node or communication failures exist.

However, there exists a large class of monitoring queries that cannot be expressed using static inclusion conditions. Examples include vehicle tracking and equipment monitoring applications where inclusion predicates need to be conditioned on readings taken by the sensor nodes such as noise levels or temperature readings. In its most simple form a dynamic inclusion predicate may be a condition of the form "current reading > threshold". More complex forms may require the evaluation of a user defined function over a history of accumulated readings. In the case of approximate evaluation of queries over the sensor data [6,18,21], the inclusion predicate is satisfied when the current sensor reading deviates by more than a given threshold from the last transmitted value. We call such predicates, whose evaluation depends also on the readings taken by the nodes, as dynamic predicates as they specify which nodes should include their response in the query evaluation at each epoch (i.e., nodes whose values exceed a given threshold, or deviate significantly from previous readings). We term those monitoring queries that contain dynamic predicates as *event monitoring queries* (EMQs).

Given a monitoring query, existing techniques seek to develop *aggregation trees* that specify the way that the data is forwarded from the sensor nodes to the `Root` node. Periodically these aggregation trees may be reorganized in order to adapt to evolving data characteristics [21].

An important characteristic of EMQs, which is not taken into account by existing algorithms that design aggregation trees, is that each sensor node may participate in the query evaluation, by including its reading in the query result, only a limited number of times, based on how often the inclusion conditions are satisfied. We can thus associate an *epoch participation frequency* P_i with each sensor node S_i, which specifies the fraction of epochs that this node participated in the query result in the recent past.

4   Problem Formulation

Our current framework supports distributive (e.g., COUNT, SUM, MAX, MIN) and algebraic (e.g., AVG) queries involving aggregate functions over the measurements collected by the participating sensor nodes. A good classification of aggregate functions is presented in [16], depending on the amount and type of state required in non-leaf nodes in order for them to calculate the aggregate result for the partition of descendant, in the aggregation tree, participating sensors. In our future work we plan to extend our framework to support all types of aggregate and non-aggregate queries.

4.1   Problem Definition

In this paper we seek to develop dissemination protocols for distributive and algebraic EMQs. The goal is, given the type of query at question, to design the aggregation tree so as to minimize either:

1. The number of transmitted messages in the network.
2. The overall energy consumption in the network.

Our algorithms do not make any assumptions about the placement of the sensor nodes, their characteristics or their radio models. However, in order to simplify the presentation, in our discussion we will focus on networks where any communication between pairs of sensor nodes is either bidirectional or impossible.

4.2   Energy Consumption Cost Model

A sensor node consumes energy at all stages of its operation. However, this energy consumption is minimal when the sensor is in a sleep mode. Furthermore, the energy drain due to computations may, in some applications, be significant, but it is typically much smaller than the cost of communication [17]. Due to this fact and because our algorithms do not require any significant computational effort by the sensor nodes, we ignore in the cost model the power consumption when the sensor node is idle and the consumption due to computations. We will thus focus on capturing the energy drain due to data communication in data driven applications. In particular, we need to estimate the energy consumption of a node S_i when either transmitting, receiving or idle listening for data. The notation that will be used in our discussion here, and later in the description of our algorithms, is presented in Table 3. Additional definitions and explanations are presented in appropriate areas of the text.

We first describe the cost model used to estimate the energy consumption of a node S_i during the data transmission of $|aggr| > 0$ bits of data to node S_j, which lies in distance $dist_{i,j}$ from S_i. The energy cost can be estimated using a linear model [20] as:

$$E_{tr_{i,j}} = SC_i + (H + |aggr|) \times (E_{TX_i} + E_{RF_i} \times dist_{i,j}^2),$$

where: (i) SC_i denotes the energy startup cost for the data transmission of S_i. This cost depends on the radio used by the sensor node; (ii) H denotes the size of the packet's header; (iii) E_{TX_i} denotes the per bit power dissipation of the transmitter electronics; and (iv) E_{RF_i} denotes the per bit and squared distance power delivered by the power amplifier. This power depends on the maximum desired communication range and, thus, from the distance of the nodes with which S_i desires to communicate. Thus, the additional energy consumption required to augment an existing packet from S_i to S_j with additional $|aggr|$ bits can be calculated as: $DE_{tr_{i,j}} = |aggr| \times (E_{TX_i} + E_{RF_i} \times dist_{i,j}^2)$.

Table 3. Symbols Used in our Algorithm

Table 2. Typical Radio Parameters

Symbol	Typical Value
SC	$1\mu J$
E_{TX}	$50nJ/bit$
E_{RF}	$100pJ/bit/m^2$
E_{RX}	$50nJ/bit$

Symbol	Description
Root	The node that initiates a query and which collects the relevant data of the sensor nodes
S_i	The i-th sensor node
P_i	The epoch participation frequency of S_i
D_i	The minimum distance, in number of hops, of S_i from the Root
$|aggr|$	The size of the aggregate values transmitted by a node
$E_{tr_{i,j}}$	Energy spent by S_i to transmit a new packet of $|aggr|$ bits to S_j
$DE_{tr_{i,j}}$	Energy spent by S_i to transmit additional $|aggr|$ bits to S_j (on an existing packet).
$AC_{i,j}$	Attachment cost of S_i to a candidate parent S_j
CF_i	Cost factor utilized by neighboring nodes of S_i when estimating their attachment cost to S_i

When a sensor node S_i receives $H + b_j$ bits from node S_j, then the energy consumed by S_i is given by: $E_{rec_i} = E_{RX_i} \times (H + b_j)$, where the value of E_{RX_i} depends on the radio model. Some typical values [20] of SC, E_{TX}, E_{RX} and E_{RF} are presented in Table 2.

The energy consumed by a sensor node when idle listening for data is significant and often comparable to the energy of receiving data. For example, in the popular MICA2 nodes the ratios for radio power draw during idle-listening, receiving of a message and transmission are 1:1:1.41 at 433MHz with RF signal power of 1mW in transmission mode [27]. Thus, due to the similar energy consumption by a sensor while either receiving or idle listening for data, our algorithms focus on the energy drain during the transmission of data.

5   Algorithm Overview

We now present our algorithms for creating and maintaining an aggregation tree that minimizes the desired metric (number of messages or energy consumption) for algebraic or distributive aggregate EMQs. Our algorithms are based on a top-down formation of the aggregation tree. The intuition behind such an approach is that the epoch participation frequency of each node in the aggregation tree influences the transmission frequency of only nodes that lie in its path to the Root. We thus demonstrate in this section that estimating the magnitude of this influence can be easily achieved by a top-down construction of the aggregation tree, while requiring the transmission of only a small set of statistics.

5.1   Construction/Update of the Aggregation Tree

The algorithm is initiated with the query propagation phase and periodically, when the aggregation tree is scheduled for reorganization. The query is propagated from the base station through the network using a flooding algorithm. In densely populated sensor networks, a node S_i may receive the announcement of the query from several of its neighbors. As in [16,26] the node will select one of these nodes as its *parent node*.

The chosen parent will be the one that exhibits the lowest *attachment cost*, meaning the lowest expected increase in the objective minimization function. For example, if our objective is to minimize the total number of transmitted messages, then the selection will be the node that is expected to result in the lowest increase in the number of transmitted messages in the *entire* path from that sensor until the Root node (and similarly for the rest of the minimization metrics). At this point we simply note that in order for other nodes to compute their attachment cost, node S_i transmits a small set of statistics $Stats_i$ and defer their exact definition for Section 5.2.

The result of this process is an aggregation tree towards the base station that initiated the flooding process. A key point in our framework is that the preliminary selection of a parent node may be revised in a second step where each node evaluates the cost of using one of its sibling nodes as an alternative parent. Due to the nature of the query propagation, and given simple synchronization protocols, such as those specified in [16], the nodes lying k hops from the Root node will receive the query announcement before the nodes that lie one hop further from the Root node. Let $RecS_k$ denote the set of nodes that receive the query announcement for the first time during the k-th step of the query propagation phase.

At step k of the query propagation phase, after the preliminary parent selection has been performed, each node S_i in set $RecS_k$, needs to consider whether it is preferable to alter its current selection and choose as its parent a *sibling node* within set $RecS_k - S_i$.[1] Each node calculates a new set of statistics $Stats_i$, based on its preliminary parent selection, and transmits an *invitation*, which also includes the node's newly calculated $Stats_i$ values, that other nodes in $RecS_k$ (and only these nodes) may accept. Of course, we need to be careful at this point and make sure that at least one node within $RecS_k$ will not accept any invitation, as this would create a disconnected network and prevent nodes from $RecS_k$ to forward their results to nodes belonging in $RecS_{k-1}$. We will achieve this by imposing a simple set of rules regarding when an invitation may be accepted by a sensor node.

Let $CandPar_i$ denote the set of nodes in $RecS_k$ that transmitted an invitation that S_i received. Let S_m be the preliminary parent node of S_i, as decided during query propagation. Amongst the nodes in $CandPar_i$, node S_i considers the node S_p such as the attachment cost $AC_{i,p}$ is minimized. If ties occur, then these are broken using the node identifiers (i.e., prefer the node with the highest id).[2] Then S_p is selected as the parent of S_i *instead of the preliminary choice* S_m only if all of the following conditions apply:

- $AC_{i,p} < AC_{i,m}$. This conditions ensures that S_p seems as a better candidate parent than the current selection S_m.
- $AC_{i,p} \leq AC_{p,i}$. This conditions ensures that it is better to select S_p as the parent of S_i, than to select S_i as the parent of S_p.

---

[1] Please note that at this step any initially selected parent of a sibling node that lies within the transmission range of S_i has already been examined in the preliminary parent selection phase and does not need to be considered.

[2] Alternative choices are equally plausible. For example, prefer the nodes with the highest/lowest identifiers depending on whether this is an odd/even invocation of the aggregation tree formation algorithm.

- If $AC_{i,p} = AC_{p,i}$, then the identifier of S_p is also larger than the identifier of S_i. This condition is useful in order to allow nodes to forward messages through neighbor nodes in $RecS_k$ and also helps break ties amongst nodes and to prevent the creation of loops.

The aggregation tree may periodically get updated because of a significant change in the data distribution. Such updates are triggered by the base station using the same protocol used in the initial creation. In this case, the nodes compute and transmit their computed statistics in the same manner, but do not need to propagate the query itself.

5.2 Calculating the Attachment Cost

Determining the candidate parent with the lowest attachment cost is not an easy decision, as it depends on several parameters. For example, it is hard to quantify the resulting transmission probability of S_j, if a node S_i decides to select S_j as its parent node. In general, the transmission frequency of S_j (please note that this is different than the epoch participation frequency of the node) may end up being as high as $\min\{P_i + P_j, 1\}$ (when nodes transmit on different epochs) and as low as P_j (when transmissions happen on the same epochs and $P_i \leq P_j$). A commonly used technique that we have adopted in our work is to consider that the epoch participation by each node is determined by independent events. Using this independence assumption, node S_j will end up transmitting with a probability $P_i + P_j - P_iP_j$, an increase of $P_i(1 - P_j)$ over P_j. Similarly, if S_{j-1} is the parent of S_j, this increase will also result in an increase in the transmission frequency of S_{j-1} by $P_i(1-P_j)(1-P_{j-1})$, etc. In our following discussion, for ease of presentation, when considering the attachment cost of S_i to a node S_j, we will assume that the nodes in the path from S_j to the Root node are the nodes $S_{j-1}, S_{j-2}, \ldots, S_1$.

Minimizing the Number of Transmissions. The attachment cost of S_i when selecting S_j as its parent node can be calculated by the increase in the transmission frequency of each link from S_i to the Root node as:

$$AC_{i,j} = P_i + P_i(1 - P_j) + P_i(1 - P_j)(1 - P_{j-1}) + \cdots$$

A significant problem concerning the above estimation of $AC_{i,j}$ is that its value depends on the epoch participation frequencies of all the nodes in the path of S_j to the Root node. Since the number of these values depends on the actual distance, in number of hops, of S_j to the Root node, such a solution does not scale in large sensor networks.

Fortunately, there exists an alternative formula to calculate the above attachment cost. Our technique is based on a recursive calculation based on a single *cost factor* CF_i at each node S_i. In our example discussed above, the values of CF_i and $AC_{i,j}$ can be easily calculated as:

$$CF_i = (1 - P_i) \times (1 + CF_j)$$
$$AC_{i,j} = P_i \times (1 + CF_j)$$

One can verify that expanding the above recursive formula and setting as the boundary condition that the CF value of the Root node is zero gives the desired result. Thus, only the cost factor, which is a single statistic, is needed at each node S_j in order for all the other nodes to be able to estimate their attachment cost to S_j.

Minimizing Total Energy Consumption, Distributive and Algebraic Aggregates
This case is very similar to the case described above. When considering the attachment cost of S_i to a candidate parent S_j, we note that additional energy is consumed by nodes in the path of S_j to the Root node only if a new transmission takes place. This is because each node aggregates the partial results transmitted by its children nodes and transmits a new single partial aggregate for its sub-tree [16]. Thus, the size of the transmitted data is independent of the number of nodes in the subtree, and only the frequency of transmission may get affected. Let $E_{tr_{i,j}}$ denote the energy consumption when S_i transmits a message to S_j consisting of a header and the desired aggregate value(s) - based on whether this is a distributive or an algebraic aggregate function. The energy consumption follows the cost model presented in Section 4.2, where the E_{RF_i} value may depend on the distance between S_i and S_j (thus, the two indices used above). Using the above notation, and similarly to the previous discussion, the attachment cost $AC_{i,j}$ is calculated as:

$$AC_{i,j} = P_i \times E_{tr_{i,j}} + P_i \times (1 - P_j) \times E_{tr_{j,j-1}} +$$
$$P_i \times (1 - P_j) \times (1 - P_{j-1}) \times E_{tr_{j-1,j-2}} + \dots$$
$$= P_i \times (E_{tr_{i,j}} + CF_j), \qquad \text{where}$$
$$CF_i = (1 - P_i) \times (E_{tr_{i,j}} + CF_j)$$

If one wishes to take the receiving cost of messages into account, all that is required is to replace in the above formulas the symbols of the form $E_{tr_{k,p}}$ with $(E_{tr_{k,p}} + E_{rec_p})$, since each message transmitted by S_k to S_p will consume energy during its reception by S_p.

A final and important note that we need to make at this point involves the estimation of the attachment cost when seeking to minimize the overall energy consumption in all the types of queries discussed in this paper. When each sensor node S_i examines the invitations of neighboring nodes (and only in this step) and estimates the attachment cost to any node S_j, in our implementation it utilizes the same E_{RF} value in order to determine the value of $E_{tr_{i,j}}$, independently on the distance of S_i to S_j. This is done so that the value of $E_{tr_{i,j}}$ is the same for all candidate parents of S_i, as desired by the proof of Theorem 1 in order to guarantee the lack of loops in the formed aggregation tree.

Theorem 1. *For sensor networks that satisfy the connectivity requirements of Section 4.1 our algorithm always creates a connected routing path that avoids loops.*

Proof: We only sketch the proof here. It is obvious that any node that will receive the query announcement will select some node as its parent node. We first demonstrate that no loops can be introduced and prove this by contradiction. Assume that the parent relationships in the created loop are as follows: $S_1 \rightarrow S_2 \rightarrow \dots S_p \rightarrow S_1$. Let D_i be the distance (in number of hops) of node S_i from the Root. Since each node can select as its parent node a node with equal or lower D value, the existence of a path from S_1 to S_2 means that $D_2 \leq D_1$ and the existence of a path from S_2 to S_1 means that $D_1 \leq D_2$. Therefore, $D_1 = D_2$.

The attachment cost $AC_{i,j}$ calculated using the aforementioned statistics is of the form: $P_i \times (a_i + CF_j)$, where a_i is a constant for each node S_i. Considering that S_1

selected S_2 as its parent and not S_p, we get: $AC_{1,2} \leq AC_{1,p} \implies CF_2 \leq CF_p$. By creating such inequalities between the current parent and child of each node, summing these up (please note that because one of the nodes in the loop will exhibit the highest identifier, for at least one of the above inequalities the equality is not possible), we get that: $CF_1 + \ldots + CF_p < CF_1 + \ldots + CF_p$. We therefore reached a contradiction, which means that our algorithm cannot create any loops.   ■

An interesting observation that we have not mentioned so far involves the nodes with zero epoch participation frequencies. For these nodes, the computed attachment costs to any neighboring node will also be zero. In such cases we select the candidate parent which produces the lowest value for the attachment cost if we ignore the node's epoch participation frequency. This decision is expected to minimize the attachment cost, if the sensor at some point starts observing events.

Minimizing Other Metrics. Our techniques can be easily adapted to incorporate additional minimization metrics. For example, the formulas for minimizing the number of transmitted bits can be derived using the formulas for the energy minimization for the corresponding type of query. In these formulas one simply has to substitute the term $E_{tr_{i,j}}$ with the size of a packet (including the packet's header) and to substitute the term $DE_{tr_{i,j}}$ with the size of each transmitted aggregate value (thus, ignoring the header size). In the case where the goal is to maximize the minimum energy amongst the sensor nodes, the attachment cost can be derived from the minimum energy, amongst the nodes in a sensor's path to the Root node, raised to -1 (since our algorithms select the candidate parent with the *minimum* attachment cost).

6   Experiments

We developed a simulator for testing the algorithms proposed in this paper under various conditions. In our discussion we term our algorithm for minimizing the number of transmissions as *MinMesg*, and our algorithm for minimizing the overall energy consumption as *MinEnergy*. Our techniques are compared against two intuitive algorithms. In the *MinHops* algorithm, each sensor node that receives the query announcement randomly selects as its parent node a sensor amongst those with the minimum distance, in number of hops, from the Root node [16]. In the *MinCost* algorithm, each sensor seeks to minimize the sum of the squared distances amongst the sensors in its path to the Root node, when selecting its parent node. Since the energy consumed by the power amplifier in many radio models depends on the square of the communication range, the *MinCost* algorithm aims at selecting paths with low communication cost.

In all sets of experiments we place the sensor nodes at random locations over a rectangular area. The radio parameters were set according to the values in Table 2. The message header was set to 32 bits, similarly to the size of each statistic and half the size of each aggregate value. In all figures we account for the overhead of transmitting statistics and invitation messages during the creation of the aggregation tree in our algorithms. All numbers presented are averages from a set of five independent experiments with different random seeds.

6.1   Experiments with Synthetic Data Sets

We initially placed 36 sensor nodes in a 300x300 area, and then scaled up to the point of having 900 sensors. We set the maximum broadcast range of each sensor to 90m. In all cases the Root node was placed on the lower left part of the sensor field. We set the epoch participation frequency of the sensor nodes with the maximum distance, in hop count, from the Root to 1. Unless specified otherwise, with probability 8% some interior node assumed an epoch participation frequency of 1, while the epoch participation frequency of the remaining interior nodes was set to 5%.

We first evaluated a SUM aggregate query over the values of epoch participating sensor nodes using all algorithms. We present the total number of transmissions for each algorithm and number of sensors in Fig. 1. The corresponding average energy consumption by the sensor nodes for each case is presented in Table 4.

As we can see, our *MinMesg* algorithm achieves a significant reduction in the number of transmitted messages compared to the MinHops and MinCost algorithms. The increase in messages induced by the *MinHops* and *MinCost* algorithms compared to our approach is up to 86% and 120%, respectively, with an average increase of 66% and 93%, respectively. However, since these gains depend on the number of transmissions that epoch-participating nodes perform, it is perhaps more interesting to measure the

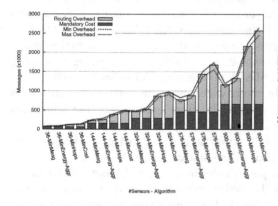
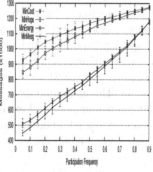

Fig. 1. Messages and Average Message Overhead for Synthetic Data Set

Fig. 2. Transmissions Varying the Epoch Participation Frequency

Table 4. Average Power Consumption (in mJ) for Synthetic Dataset with Error Bounds

Table 5. AveragePowerConsumption(in mJ) for SchoolBuses Dataset with Error Bounds

Sensors	Aggregate SUM Query			
	MinMesg	MinEnergy	MinHops	MinCost
36	94.38	87.20	166.12	125.78
144	72.84	67.56	140.81	117.18
324	66.06	61.83	141.46	103.22
576	62.52	58.73	133.71	101.84
900	61.29	56.68	127.79	99.93
±	7.51%	10.94%	5.66%	6.8%

# Sensors	MinMesg	MinEnergy	MinHops	MinCost
150	75.65	67.64	94.33	75.51
600	61.19	51.10	84.67	58.74
1350	58.87	47.86	85.89	55.48
±	9.58%	5.5%	15.01%	17.77%

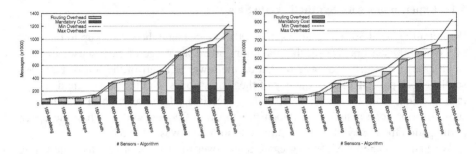

Fig. 3. Transmissions - Trucks data   **Fig. 4.** Transmissions - SchoolBuses data

routing overhead of each technique. We define the routing overhead of each algorithm as the relative increase in the number of transmissions when compared to the number of epoch participations by the sensor nodes. Note that the latter number is a *mandatory* cost that represents the transmissions in the network if each sensor could communicate directly with the Root node. For example, if the total number of epoch participations by the sensor nodes was 1000, but the overall number of transmissions was 1700, then the routing overhead would have been equal to $(1700 - 1000)/1000 = 70\%$. As we observe from Fig. 1, our *MinMesg* algorithm often results in 3 times smaller routing overhead compared to the alternative algorithms considered. We also observe that the MinEnergy algorithm in the aggregate case produced results very close to the ones of MinMesg. A main difference between these two algorithms is that amongst candidate parents with similar cost factors, the MinEnergy algorithm is less likely to select a distant neighbor than the MinMesg algorithm, which only considers epoch participation frequencies. This is a trend that we observed in all our experiments. The MinEnergy algorithm performs very well in this experiment. Compared to the MinHops algorithm, it achieves up to a 2-fold reduction in the power drain. Compared to the MinCost algorithm the energy savings are smaller but still significant (i.e., up to 76%).

We expect that the more the epoch participation frequencies of sensor nodes increase, the less likely that out techniques will be able to provide substantial savings compared to the MinHops and MinCost algorithms. In Fig. 2 we repeat the aggregate query of Fig. 1 at the sensor network with 324 nodes, but vary the epoch participation frequency P_i of those nodes that do not make a transmission at each epoch (i.e., of those nodes with $P_i < 1$). While Fig. 2 validates our intuition, it also demonstrates that significant savings can be achieved even when sensor nodes have large P_i values (i.e., $P_i \geq 0.5$).

6.2   Experiments with Real Data Sets

We also experimented with the following two real data sets. The **Trucks** data set contains trajectories of 276 moving trucks [1]. Similarly, the **SchoolBuses** data set contains trajectories of 145 moving schoolbuses [1]. For each data set we initially overlaid a sensor network of 150 nodes over the monitored area. We set the broadcast range such that interior sensor nodes could communicate with at least 5 more sensor nodes. Moreover, each sensor could detect objects within a circle centered at the node and with radius equal to 60% of the broadcast range. We then scaled the data set up to a network of 1350 sensors, while keeping the sensing range steady. In Figs. 3 and 4 we depict the total number of transmissions by all algorithms for the Trucks and SchoolBuses data sets, correspondingly, when computing the SUM of the number of detected objects. In our scenario, nodes that do not observe an event make a transmission only if they need to propagate measurements/aggregates by descendant nodes. We present the average energy consumption of the sensor nodes in the same experiment for the SchoolBuses data set in Table 5. As it is evident, our algorithms achieve significant savings in both metrics. For example, the MinCost algorithm, which exhibits lower power consumption than the MinHops algorithm, still drains about 15% more energy than our MinEnergy algorithm. Moreover, both our MinMesg and MinEnergy algorithms significantly reduce the amount of transmitted messages by up to 31% and 53% when compared to the MinHops and MinCost algorithms, respectively.

7 Conclusions

In this paper we presented algorithms for building and maintaining efficient aggregation trees in support of event monitoring queries in wireless sensor networks. We demonstrated that is it possible to create efficient aggregation trees that minimize important network resources using a small set of statistics that are communicated in a localized manner during the construction of the tree. Furthermore, our techniques utilize a novel 2-step refinement process that significantly increases the quality of the created trees. In our future work, we plan to extend our framework to also support holistic aggregates and SELECT * queries, as well as extend our framework for a multi-query setting.

References

1. Rtree Portal, http://www.rtreeportal.org
2. Abadi, D.J., Madden, S., Lindenr, W.: REED: Robust, Efficient Filtering and Event Detection in Sensor Networks. In: VLDB (2005)
3. Cerpa, A., Estrin, D.: ASCENT: Adaptive Self-Configuring sEnsor Network Topologies. In: INFOCOM (2002)
4. Chang, J.-H., Tassiulas, L.: Energy Conserving Routing in Wireless Ad-hoc Networks. In: INFOCOM (2000)
5. Considine, J., Li, F., Kollios, G., Byers, J.: Approximate Aggregation Techniques for Sensor Databases. In: ICDE (2004)
6. Deligiannakis, A., Kotidis, Y., Roussopoulos, N.: Hierarchical In-Network Data Aggregation with Quality Guarantees. In: Bertino, E., Christodoulakis, S., Plexousakis, D., Christophides, V., Koubarakis, M., Böhm, K., Ferrari, E. (eds.) EDBT 2004. LNCS, vol. 2992, pp. 658–675. Springer, Heidelberg (2004)
7. Deligiannakis, A., Kotidis, Y., Roussopoulos, N.: Dissemination of Compressed Historical Information in Sensor Networks. VLDB Journal (2007)
8. Demers, A., Gehrke, J., Rajaraman, R., Trigoni, N., Yao, Y.: The Cougar Project: A Work In Progress Report. SIGMOD Record 32(4), 53–59 (2003)
9. Deshpande, A., Guestrin, C., Madden, S., Hellerstein, J.M., Hong, W.: Model-Driven Data Acquisition in Sensor Networks. In: VLDB (2004)

10. Duckham, M., Nittel, S., Worboys, M.: Monitoring Dynamic Spatial Fields Using Responsive Geosensor Networks. In: GIS (2005)
11. Estrin, D., Govindan, R., Heidermann, J., Kumar, S.: Next Century Challenges: Scalable Coordination in Sensor Networks. In: MobiCOM (1999)
12. Gao, J., Guibas, L.J., Milosavljevic, N., Hershberger, J.: Sparse Data Aggregation in Sensor Networks. In: IPSN (2007)
13. Intanagonwiwat, C., Estrin, D., Govindan, R., Heidermann, J.: Impact of Network Density on Data Aggregation in Wireless Sensor Networks. In: ICDCS (2002)
14. Kotidis, Y.: Snapshot Queries: Towards Data-Centric Sensor Networks. In: ICDE (2005)
15. Kotidis, Y.: Processing Proximity Queries in Sensor Networks. In: Proceedings of the 3rd International VLDB Workshop on Data Management for Sensor Networks (DMSN) (2006)
16. Madden, S., Franklin, M.J., Hellerstein, J.M., Hong, W.: Tag: A Tiny Aggregation Service for ad hoc Sensor Networks. In: OSDI Conf. (2002)
17. Madden, S., Franklin, M.J., Hellerstein, J.M., Hong, W.: The Design of an Acquisitional Query processor for Sensor Networks. In: ACM SIGMOD (2003)
18. Olston, C., Widom, J.: Offering a Precision-Performance Tradeoff for Aggregation Queries over Replicated Data. In: VLDB (2000)
19. Pattem, S., Krishnamachari, B., Govindan, R.: The Impact of Spatial Correlation on Routing with Compression in Wireless Sensor Networks. In: IPSN (2004)
20. Raghunathan, V., Schurgers, C., Park, S., Srivastava, M.: Energy aware wireless microsensor networks. IEEE Signal Processing Magazine 19(2) (2002)
21. Sharaf, A., Beaver, J., Labrinidis, A., Chrysanthis, P.: Balancing Energy Efficiency and Quality of Aggregate Data in Sensor Networks. VLDB Journal (2004)
22. Silberstein, A., Braynard, R., Yang, J.: Constraint Chaining: On Energy Efficient Continuous Monitoring in Sensor Networks. In: SIGMOD (2006)
23. Singh, S., Woo, M., Raghavendra, C.S.: Power-aware routing in mobile ad hoc networks. In: ACM/IEEE International Conference on Mobile Computing and Networking (1998)
24. Trigoni, N., Yao, Y., Demers, A.J., Gehrke, J., Rajaraman, R.: Multi-query Optimization for Sensor Networks. In: Prasanna, V.K., Iyengar, S.S., Spirakis, P.G., Welsh, M. (eds.) DCOSS 2005. LNCS, vol. 3560, pp. 307–321. Springer, Heidelberg (2005)
25. Xue, W., Luo, Q., Chen, L., Liu, Y.: Contour Map Matching for Event Detection in Sensor Networks. In: SIGMOD (2006)
26. Yao, Y., Gehrke, J.: The Cougar Approach to In-Network Query Processing in Sensor Networks. SIGMOD Record 31(3), 9–18 (2002)
27. Ye, W., Heidermann, J.: Medium Access Control in Wireless Sensor Networks. Technical report, USC/ISI (2003)
28. Zeinalipour-Yazti, D., Andreou, P., Chrysanthis, P.K., Samaras, G., Pitsillides, A.: The Micropulse Framework for Adaptive Waking Windows in Sensor Networks. In: MDM, pp. 351–355 (2007)

Improving Chord Network Performance Using Geographic Coordinates

Sonia Gaied Fantar[1] and Habib Youssef[2]

Research Unit Prince, ISITC Hammam Sousse, University of Sousse, Tunisia
soniagaied3@gmail.com
habib.youssef@fsm.rnu.tn

Abstract. Structured peer-to-peer overlay networks such as Chord, CAN, Tapestry, and Pastry, operate as distributed hash tables (DHTs). However, since every node is assigned a unique identifier in the basic design of DHT (randomly hashed), "locality-awareness" is not inherent due to the topology mismatching between the P2P overlay network and the physical underlying network. In this paper, we propose to incorporate physical locality into a Chord system. To potentially benefit from some level of knowledge about the relative proximity between peers, a network positioning model is necessary for capturing physical location information of network nodes. Thus, we incorporate GNP (Global Network Positioning) into Chord (Chord-GNP) since peers can easily maintain geometric coordinates that characterize their locations in the Internet. Next, we identify and explore three factors affecting Chord-GNP performance: distance between peers, message timeout calculation and lookup latency. The measured results show that Chord-GNP efficiently locates the nearest available node providing a locality property. In addition, both the number of the messages necessary to maintain routing information and the time taken to retrieve data in Chord-GNP is less than that in Chord.

1 Introduction

A large number of structured Peer-To-Peer overlay systems constructed on top of DHT such as Chord [1], CAN [2], Tapestry [4] or Pastry [3] have been proposed recently. Due to their scalability, robustness and self-organizing nature, these systems provide a very promising platform for a range of large-scale and distributed applications.

In the structured P2P model, the nodes in the network, called peers, form an application-level overlay network over the physical network. This means that the overlay organizes the peers in a network in a logical way so that each peer connects to the overlay network just through its neighbors. However, the mechanism of peers randomly choosing logical neighbors without any knowledge about underlying physical topology can cause a serious topology mismatching between the P2P overlay network and the physical underlying network. The topology mismatching problem brings a great stress in the Internet infrastructure and greatly

N. Trigoni, A. Markham, and S. Nawaz (Eds.): GSN 2009, LNCS 5659, pp. 77–86, 2009.

limits the performance gain from various lookup or routing techniques. Meanwhile, due to the inefficient overlay topology and the absence of relationship between the node's location and the node's identifier, the DHT-based lookup mechanisms cause a large volume of unnecessary traffic. Moreover, due to this discrepancy, DHTs do not offer a guarantee on the number of physical hops taken during a lookup process, as a single overlay hop is likely to involve multiple physical routing hops. Aiming at alleviating the mismatching problem and reducing the unnecessary traffic on Chord, we propose a novel location-aware identifier assignment function. This function attributes identifiers to nodes by choosing physically closer nodes as logical neighbors. Thus, the main contribution of this paper is the design and analysis of a new approach that incorporates locality-awareness into Chord identifiers, by assigning identifiers to nodes that reflects their geographic disposition. So, we propose to use a coordinates-based mechanism, called Global Network Positioning (GNP), to predict Internet network distance. Chord-GNP is constructed on the basis of Chord aiming to achieve better routing efficiency, which attributes peer's identifiers by choosing physically closer nodes as logical neighbors. The main optimizations in Chord-GNP are lower overlay hops and lookup latency.

The remainder of this paper is organized as follows.

We discuss related work in Section 2. In Section 3, we present Chord in detail. Section 4 talks about the Global Network Positioning (GNP). In sections 5 and 6 we analyze our contribution and give simulation results. Section 7 concludes this paper and gives a brief outlook on our future work.

2   Related Work

Many efforts have been made to improve locality awareness in decentralized structured peer-to-peer overlays. The most widely used approaches in locality awareness are network proximity. Castro et al [5] divide techniques to exploit network proximity into three categories: expanding-ring search, heuristics, and landmark clustering. The entries of routing table are chosen as the topologically nearest among all nodes with node's identifier in the desired portion of the key space [6]. The success of this technique depends on the degree of freedom an overlay protocol has in choosing routing table entries without affecting the expected number of routing hops. Another limitation of this technique is that it does not work for overlay protocols like CAN and Chord, which require that routing table entries refer to specific points in the key space.

3   CHORD

Chord is a peer-to-peer protocol which presents a new approach to the problem of efficient location. Chord [1] uses consistent hashing [7] to assign keys to its peers. Consistent hashing is designed to let peers enter and leave the network with minimal interruption. This decentralized scheme tends to balance the load on the system, since each peer receives roughly the same number of keys, and there is little movement of keys when peers join and leave the system. In a steady state, for N peers in the system, each peer maintains routing state information for about only $O(logN)$ other peers (N number of peers in the system). The consistent hash functions assign peers and data keys an m-bit identifier using SHA-1 [8] as the base hash function. A peer's identifier is chosen by hashing the peer's IP address, while a key identifier is produced by hashing the data key. The length of the identifier m must be large enough to make the probability of keys hashing to the same identifier negligible. Identifiers are ordered on an identifier circle modulo $2m$. Key k is assigned to the first peer whose identifier is equal to or follows k in the identifier space. This peer is called the successor peer of key k, denoted by successor(k). If identifiers are represented as a circle of numbers from 0 to 2^{m-1}, then successor(k) is the first peer clockwise from k. The identifier circle is termed as the Chord ring.

To maintain consistent hashing mapping when a peer n joins the network, certain keys previously assigned to n's successor now need to be reassigned to n.

4 Predicting Internet Network Distances with Coordinates Based Approaches

Among several categories of approaches that predict internet network distance, the coordinates based approaches may be the most promising. Several approaches have been proposed among which GNP [9][10] may have received the most attention.

4.1 Global Network Positioning (GNP)

GNP [9][10] is a two-part architecture that is proposed to enable the scalable computation of geometric host coordinates in the Internet. In the first part, a small distributed set of hosts called Landmarks first compute their own coordinates in a chosen geometric space. The Landmarks' coordinates serve as a frame of reference and are disseminated to any host who wants to participate. In the second part, equipped with the Landmarks' coordinates, any end host can compute its own coordinates relative to those of the Landmarks.

Landmark Operations. The first part of the architecture is to use a small distributed set of hosts known as Landmarks to provide a set of reference coordinates necessary to orient other hosts. When a re-computation of Landmarks' coordinates is needed over time, we can ensure the coordinates are not drastically changed if we simply input the old coordinates instead of random numbers as the start state of the minimization problem. Once the Landmarks' coordinates are computed, they are disseminated to any ordinary host that wants to participate in GNP.

Ordinary Host Operations. In the second part of our architecture, ordinary hosts are required to actively participate. Using the coordinates of the Landmarks in a geometric space, each ordinary host now derives its own coordinates.

To do so, an ordinary host H measures its round-trip times to the N Landmarks using ICMP ping messages and takes the minimum of several measurements for each path as the distance. Using the measured host-to-Landmark distances, host H can compute its own coordinates that minimize the overall error between the measured and the computed host-to-Landmark distances.

5   Proposition

One important issue in alleviating the mismatching problem and reducing the unnecessary traffic on Chord is to attribute peer's identifiers by choosing physically closer nodes as logical neighbors. Thus, we can benefit from predicting internet network distances. Specifically, we propose to use a coordinates-based approaches for network distance prediction in Chord architecture. The main idea is to ask peers to maintain coordinates (i.e. a set of numbers) that characterize their locations in the Internet such that network distances can be predicted by evaluating a distance function over hosts' coordinates. Our contribution is to allow participating peers in Chord to collaboratively construct an overlay based on physical location. And, to potentially benefit from some level of knowledge about the relative proximity between the peers, the Global network Positioning approach (GNP) [9][10] is integrated into Chord for capturing physical location information of network peers. This paper presents a topology-based node identifier assignment for Chord that attempts to map the overlay's logical key space onto the physical network such that neighboring nodes in the key space are close in the physical network.

5.1   Descriptions

Using the Chord lookup protocol, the peers are assumed to be distributed uniformly at random on the ring. In particular, there is a base hash function which maps peers, based on their IP adresses, to points on the circle, i.e. that it maps identifiers to essentially random locations on the ring (Figure 1.a). However, in Chord [1], when a node joins the network, it is not optimally positioned in the ring in respect of the underlying network such as IP network. For this purpose, we propose a novel location-aware identifier assignation function for Chord. This function attributes identifiers to nodes by choosing physically closer nodes as logical neighbors by using a coordinates-based mechanism (Figure 1.b). Chord could potentially benefit from some level of knowledge about the relative proximity between its participating nodes by using the coordinates based approaches GNP to predict internet network distance : Chord-GNP.

5.2   Location-Aware Identifier Assignation Function for Chord

In this section, we describe a design technique whose primary goal is to reduce the latency of Chord routing. Not unintentionally, this technique offers the additional advantage of improved Chord robustness in term of routing.

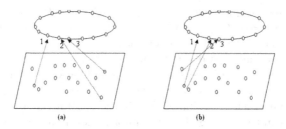

Fig. 1. (a) mapping nodes to identifiers with consistent hashing. (b) mapping nodes to identifiers with locality aware.

In our representation, we model the Internet as a particular geometric space S. Let us denote the coordinates of a host H in S as C_H^S. Then, each peer in the Internet is characterized by its position in the geometric space with a set of geometric coordinates C_H^S.

We want to replace the Chord's base hash function SHA-1 (Node identifier = SHA-1(IP address, port number)) by the location-aware identifier assignation function that generates geometric coordinate's identifiers: GNP(IP address, port number)=C_H^S

$$id_N = C_N^S \qquad (1)$$

5.3   Design

Node joining and leaving in Chord-GNP is handled like the Chord does. The difference is that in Chord, a node's identifier is chosen by hashing the node's IP address, but Chord-GNP provides location-aware identifier assignation function. This function assign peer's identifier designating its position in the virtual ring. We will describe the three most basic pieces of our design: Chord-GNP routing, construction of the Chord-GNP coordinate overlay, and maintenance of the Chord-GNP overlay.

Routing in Chord-GNP. Chord-GNP also uses a one-dimensional circular key space. Chord-GNP's main modification to Chord is to include new identifiers into Chord's routing tables, i.e. Chord-GNP inherits Chord's successor list and finger table to use in Chord-GNP's routing algorithm and maintenance algorithm. Each node has a successor list of nodes that immediately follow it in the coordinate's space.

Each node keeps a list of successors: If a node's successor fails, it advances through this list until a live node is found. Routing efficiency is achieved with the finger list of nodes spaced exponentially around the coordinate's space.

Intuitively, routing in Chord-GNP network works by following the finger table through the ring from source to destination coordinates.

Chord-GNP construction. To allow the Chord-GNP to grow incrementally, a new node that joins the system must derives its own coordinates that characterize its location in the Internet to be allocated in the ring. This is done by the coordinates based approaches: GNP. In GNP, the Internet is modeled as a D-dimensional geometric space. Peers maintain absolute coordinates in this geometric space to characterize their locations on the Internet. Network distances are predicted by evaluating a distance function over peers' coordinates.

A small distributed set of peers known as Landmarks provide a set of reference coordinates. peers measure their latencies to a fixed set of Landmark nodes in order to compute their coordinates. While the absolute coordinates provide a scalable mechanism to exchange location information in a peer-to-peer environment, the GNP scheme presented in Section 4 used distance measurements to a fixed set of Landmarks to build the geometric model. When a node joins the Chord-GNP network it will be placed in the ring by choosing physically closer nodes as logical neighbors. The successor pointers of some nodes will have to change. It is important that the successor pointers are up to date at any time because the correctness of lookups is not guaranteed otherwise. The Chord-GNP protocol uses a stabilization protocol running periodically in the background to update the successor pointers and the entries in the finger table.

Node departure, recovery and Chord-GNP maintenance. To leave an established Chord-GNP ring, a node can give its keys to its successor and then inform its predecessor. The successor and predecessor nodes then update their fingers tables and successors lists. Chord-GNP ensures also that each node's successor list is up to date. It does this using a "stabilization" protocol that each node runs periodically in the background and which updates Chord's finger tables and successor pointers. In Chord-GNP, when the successor node does not respond or fails, the node simply contacts the next node on its successor list.

6   Simulation Results

In this section, we evaluate the performance of Chord-GNP through simulation. The goal of this evaluation is to validate the proposition of Section 5. That proposition assumed idealized models of Chord.

We are implementing optimizations of GNP coordinates in the current open source implementation of the Chord distributed hash table as described in [1] Overlay Weaver [11][12]. We modified the Chord simulator [11] to implement GNP.

6.1   Distance between Peers

In an experiment, we first bring up a network of four nodes placed in 2 classrooms (figure 2). 1) Chord: Figure 3 shows screenshots of simulation which visualizes communication between nodes just in time. Nodes are sorted on the ring on the basis of their ID, taking into account the Hash algorithm as identifier assignation function. Each node has to maintain a virtual link to its successor, which is the node directly following it in the ordered node set. With this structure, any node can route messages to any other node simply by each intermediate node forwarding the message to its successor until the destination is reached. This, however,

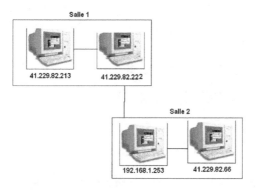

Fig. 2. Arrangement of Peers

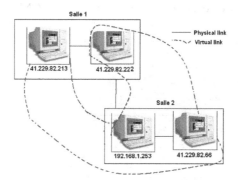

Fig. 3. Identifier assignation function is Hash algorithm SHA-1

results in choosing of the next peers, rendering routing not scalable. As was mentioned in Figure 5, node 41.229.82.222 routes messages to node 41.229.82.213 by intermediate node 41.229.82.66, not knowing that node 41.229.82.213 is closer to node 41.229.82.222.

2) Chord-GNP: In this section we report the results of testing Chord using identifiers provided by GNP.

In this experiment, Nodes are sorted on the ring on the basis of their coordinates given by GNP (Figure 4).

With this structure, node 41.229.82.222 can route messages directly to its successor which is 41.229.82.213.

6.2   Message Timeout Calculation

The performance penalty of routing in the overlay over taking the shortest path in the underlying network is quantified by the message timeout calculation. Chord sends messages periodically to maintain overlays. On a real network, we conducted experiments with four computers. We invoked a DHT shell on each

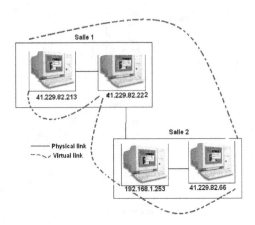

Fig. 4. Identifier assignation function is GNP

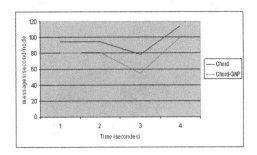

Fig. 5. Number of messages per second per node being passed between four real computers

computer and controlled them via a network. We counted the numbers using the message counter. The control scenario was as follows: after 4 nodes had joined an overlay, a node put a value on a DHT every 2 seconds 4 times. Figure 5 plot the average number of messages per second per node for Chord and Chord-GNP. We claim that the number of messages per second per node being passed between Chord-GNP's nodes are less those sent over Chord's nodes. These experiments show that Chord-GNP reduces the number of messages for Chord. In fact, the reducing of the number of messages is improved by the choice of physically closer nodes as logical neighbors.

6.3   End-to-End Path Latency

This section presents latency measurements obtained from implementations of Chord and Chord-GNP. We produce several lookup scenarios with the number of nodes increased from 10 to 300 nodes. Then, we generate a simulation topology for these scenarios and evaluate the end-to-end path lookup latencies. Figure 6

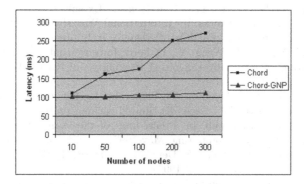

Fig. 6. Lookup latency with the number of nodes increased

shows that only where the number of nodes is small, the latency in both systems is the same. In addition, the lookup latency of Chord increases as the number of nodes is increased, while the latency of Chord-GNP remains almost at the same low level. It is an important advantage of Chord-GNP that its performance is independent of the number of participating nodes, which confirms our prediction. Consequently, the performance of Chord-GNP becomes much better than that of Chord.

7   Conclusions

In this paper, we have studied a new class of solutions to the Internet distance prediction problem that is based on end hosts-maintained coordinates, called Global Network Positioning (GNP). We have proposed to apply this solution in the context of a peer-to-peer architecture, precisely in the Chord Algorithm. This topology-based node identifier assignment attempts to map the overlay's logical key space onto the physical network such that neighboring nodes in the key space are close in the physical network. Chord-GNP's main routing optimizations are of less overlay hops and passing proximity links of the underlay network. Meanwhile, Chord-GNP has insignificant lookup latency in comparison to Chord.

References

1. Stoica, I., Morris, R., Karger, D., Kaashock, M., Balakrishman, H.: Chord: A scalable P2P lookup protocol for Internet applications. In: Proc. of ACM SIGCOMM (2001)
2. Ratnasamy, S., Francis, P., Handley, M., Karp, R., Shenker, S.: A scalable content addressable network. In: Proc. of ACM SIGCOMM (August 2001)
3. Rowstron, A., Druschel, P.: Pastry: Scalable, decentralized object location and routing for large-scale p2p systems. In: Proc. of IFIP/ACM Middleware (2001)
4. Zhao, B., Huang, L., Stribling, J., Rhea, S.C., Joseph, A., Kubiatowicz, J.: Tapestry: A global-scale overlay for rapid service deployment. IEEE J-SAC 22(1) (2004)

5. Castro, M., Druschel, P., Hu, Y.C., Rowstron, A.: Exploiting Network Proximity in Peer-to-Peer Overlay Networks. In: International Workshop on Future Directions in Distributed Computing (FuDiCo), Bertinoro, Italy (June 2002)
6. Hong, F., Li, M., Yu, J., Wang, Y.: PChord: Improvement on Chord to Achieve Better Routing Efficiency by Exploiting Proximity. In: ICDCS Workshops 2005, pp. 806–811 (2005)
7. Karger, D., Lehman, E., Leighton, T., Panigrahy, R., Levine, M., Lewin, D.: Consistent hashing and random trees: distributed caching protocols for relieving hot spots on the world wide web. In: Proceedings of the twenty-ninth annual ACM symposium on Theory of computing, May 1997, pp. 654–663 (1997)
8. Secure hash standard, NIST, U.S. Dept. of Commerce, National Technical Information Service FIPS 180-1 (April 1995)
9. Ng, T.S.E., Zhang, H.: Predicting internet network distance with coordinates-based approaches. In: Proceedings of IEEE Infocom (May 2002)
10. Ng, T.S.E., Zhang, H.: Towards Global Network Positioning. In: ACM SIGCOMM Internet Measurement Workshop, San Francisco, CA (November 2001)
11. Overlay Weaver: an overlay construction toolkit, http://overlayweaver.sf.net/
12. Shudo, K., Tanaka, Y., Sekiguchi, S.: Overlay Weaver: An overlay construction toolkit Computer Communications (Special Issue on Foundations of Peer-to-Peer Computing) 31(2), 402–412 (2008) (available online on August 14, 2007)

RFID Data Aggregation*

Dritan Bleco and Yannis Kotidis

Department of Informatics
Athens University of Economics and Business
76 Patission Street, Athens, Greece
{dritan,kotidis}@aueb.gr

Abstract. Radio frequency identification (RFID) technology is gaining popularity for many IT related applications. Nevertheless, an immediate adoption of RFID solutions by the existing IT infrastructure is a formidable task because of the volume of data that can be collected in a large-scale deployment of RFIDs. In this paper we present algorithms for temporal and spatial aggregation of RFID data streams, as a means to reduce their volume in an application controllable manner. We propose algorithms of increased complexity that can aggregate the temporal records indicating the presence of an RFID tag using an application-defined storage upper bound. We further present complementary techniques that exploit the spatial correlations among RFID tags. Our methods detect multiple tags that are moved as a group and replace them with a surrogate group ID, in order to further reduce the size of the representation. We provide an experimental study using real RFID traces and demonstrate the effectiveness of our methods.

1 Introduction

Radio frequency identification (RFID) technology has gained significant attention in the past few years. In a nutshell, RFIDs allow us to sense and identify objects. RFIDs are by no means a new technology. Its origins can be traced back to World War II, where it was deployed in order to distinguish between friendly and enemy war planes [1]. Since then, RFIDs have seamlessly infiltrated our daily activities. In many cities around the word, RFIDs are used for toll collection, in roads, subways and public buses. Airport baggage handling and patient monitoring are further examples denoting the widespread adoption of RFIDs.

With their prices already in the range of a few cents, RFID tags are becoming a viable alternative to bar codes for retail industries. Large department stores like the Metro Group and Wal-Mart are pioneers in deploying RFID tags in their supply chain [2]. Individual products, pallets and containers are increasingly tagged with RFIDs. At the same time, RFID readers, are placed at warehouse entrances, rooms and distribution hubs. These readers compute and communicate the list

---

* This work has been supported by the Basic Research Funding Program, Athens University of Economics and Business.

N. Trigoni, A. Markham, and S. Nawaz (Eds.): GSN 2009, LNCS 5659, pp. 87–101, 2009.
© Springer-Verlag Berlin Heidelberg 2009

of RFID tags sensed in their vicinity to a central station for further processing and archiving. The ability to automatically identify objects, without contact, through their RFID tags, allows for a much more efficient tracking in the supply chain, thus eliminating the need for human intervention (which for instance is typically required in the case of bar codes). This removal of latency between the appearance of an object at a certain location and its identification allows us to consider new large- or global- scale monitoring infrastructures, enabling more efficient planning and management of resources.

Nevertheless, an immediate adoption of RFID technology by existing IT infrastructure, consisting of systems such as enterprise resource planning, manufacturing execution, or supply chain management, is a formidable task. As an example, the typical architecture of a centralized data warehouse, used by decision support applications, assumes a periodic refresh schedule [3] that contradicts the need for currency by a supply chain management solution: when a product arrives at a distribution hub, it needs to be processed as quickly as possible. Moreover, existing systems have not been designed to cope with the voluminous data feeds that can be easily generated through a wide-use of RFID technology. A pallet of a few hundred products tagged with RFIDs generates hundreds of readings every time it is located within the sensing radius of a reader. A container with several hundred pallets throws tens of thousands of such readings. Moreover, these readings are continuous: the RFID reader will continuously report all tags that it senses at every time epoch. Obviously, some form of data reduction is required in order to manage these excessive volumes of data.

Fortunately, the type of data feeds generated by RFIDs are embedded with lots of redundancy. As an example, successive observations of the same tag by a reader can be easily encoded using a time interval indicating the starting and ending time of the observation. Unfortunately, this straightforward data representation is prone to data collection errors. Existing RFID deployments, routinely drop a significant amount of the tag-readings; often as much as 30% of the observations are lost [4]. This makes the previous solution practically ineffectual as it can not limit in a application-controllable manner the number of records required in order to represent an existing RFID data stream. In this paper, we investigate data reduction methods that can reduce the size of the RFID data streams into a manageable representation that can then be fed into existing data processing and archiving infrastructures such as a data warehouse. Key to our framework is the decision to move much of the processing near the locations where RFID streams are produced. This reduces network congestion and allows for large scale deployment of the monitoring infrastructure.

Our methods exploit the inherent temporal redundancy of RFID data streams. While an RFID tag remains at a certain location, its presence is recorded multiple times by the readers nearby. Based on this observation we propose algorithms of increased complexity that can aggregate the records indicating the presence of this tag using an application-defined storage upper bound. During this process some information might be lost resulting in *false positive* or *false negative* cases of identification. Our techniques minimize the inaccuracy of the reduced representation for a target space constraint. In addition to temporal correlations, RFID data streams exhibit spatial correlations as well. Packaged products within a pallet are read all together when near an RFID reader. This observation can be exploited by introducing a data representation that *groups* multiple RFID readings within the same record. While this observation has already been discussed in the literature [5], to our knowledge we are the first to propose a systematic method that can automatically identify and use such spatial correlations.

The contributions of our work are:

- We propose a distributed framework for managing voluminous streams of RFID data in an supply-chain management system. Our methods push the logic required for reducing the size of the streams at the so-called Edgeware, near the RFID readers, in an attempt to reduce network congestion.
- We present a lossy aggregation scheme that exploits the temporal correlations in RFID data streams. For a given space constraint, our techniques compute the optimal temporal representation of the RFID data stream that reduces the expected error of the approximate representation, compared to the full, unaggregated data stream. We also consider alternative greedy algorithms that produce a near-optimal representation, at a fraction of the time required by the optimal algorithm.
- We present complementary techniques that further exploit the spatial correlations among RFID tags. Our methods detect multiple tags that are moved as a group and replace them with a surrogate group ID, in order to further reduce the size of the representation.
- We provide an experimental evaluation of our techniques and algorithms using real RFID data traces. Our experiments demonstrate the utility and effectiveness of our proposed algorithms, in reducing the volume of the RFID data, by exploiting correlations both in time and space.

The rest of the paper is organized as follows. In Section 2 we discuss related work. In Section 3 we introduce the system architecture we consider in this work, present the details of an RFID data stream and state our optimization problem. In Section 4 we present our algorithms for temporal aggregation of the RFID streams, while in Section 5 we describe our spatial aggregation process. Our experimental evaluation is presented in Section 6. Finally, Section 7 contains concluding remarks.

2   Related Work

There have been several recent proposals discussing RFID technology. These works analyze RFID systems from different points of view, including hardware [6], software [2], data processing [7,8,9] and privacy [10]. The work in [11] presents an RFID system deployed inside a building where the tagged participants walking through it produce a large amount of RFID data. The authors discuss the system, its performance, showcase analysis of higher-level information inferred from raw RFID data and comment on additional challenges, such as privacy.

The main characteristics of an RFID system, such as their temporal and dynamic behavior, the inaccuracy of data, the need for integration with existing IT systems, the streaming nature of the raw data and their large volumes are discussed in [12]. The paper presents a temporal data model, DRER, which exploits the specific fundamentals of an RFID application and of the primitive RFID data. This work shows the basic features that should be included in a RFID Middleware system, such as including effective query support and automatic data acquisition and transformation. The work in [9] introduces a deferred RFID data cleaning framework of using rules executed at query time.

The work in [13] demonstrates the significance of compression in RFID systems and discusses a graph-based model that captures possible object locations and their containment relationships. However, in order to provide accurate results, the graph models require high detection rates at the RFID readers. In [4], the authors highlight the inherent unreliability of RFID data streams. Software running at the Edgeware typically corrects for dropped readings using a temporal smoothing filter based on a pre-defined sliding window over the reader's data stream that interpolates for lost readings from each tag within the time window. The work in [4] uses a statistical sampling-based approach that results in an adaptive smoothing filter, which determines the window-size automatically based on observed readings. This work nicely complements our techniques, as it may be used at a pre-processing step in order to clean the incoming RFID data stream, before applying our aggregation algorithms.

Our temporal aggregation process works by first transforming the individual readings produced by the readers into temporal segments and then reduces the number of segments while trying to minimize the error of the approximate temporal representation. At an abstract level, this process resembles the construction of a one-dimensional histogram on the frequency distribution of a data attribute, used in traditional database management systems [14,15,16]. Application of existing data compression algorithms such as wavelets [17,18], their probabilistic counterparts [19], or even algorithms developed for sensory data [20,21] on RFID data streams is an interesting research topic.

A model for data warehousing RFID data has been proposed in [5]. This work studies the movement of products from suppliers to points of sale taking advantage of bulky object movements, of data generalization and the merge or collapse of path segment that RFID objects follow. The authors introduce a basic RFID data compression scheme, based on the observation that tags move together in any stage of the movement path. However, there is no provision for missing or erroneous data tuples. The work in [22] introduced the Flowcube, a data cube computed for a large collection of paths. The Flowcube model analyzes item flows in an RFID system. The Flowcube differs from the traditional data cube [23] in that it does not compute aggregated measurements but, instead, movement trends of each specific item. The work in [24] introduced the notion of a service provisioning data warehouse, which organizes records produced by a service delivery process (such as a delivery network). The paper introduces pair-wise aggregate queries as a means to analyze massive datasets. Such a query consists of

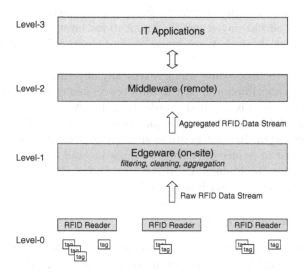

Fig. 1. System Overview

a path expression over the delivery graph and a user defined aggregate function that consolidates the recorded data. Our RFID data aggregation framework can be used for the instrumentation of a large-scale ETL process while building such a data warehouse using data emanating from RFID readers along the delivery network.

3 Preliminaries

In this section we first present an overview of the architecture we assume for managing RFID data. We then discuss in more detail the contents of an RFID data stream and a simple relational mapping that exploits temporal correlations in the stream.

3.1 System Architecture

In our work we assume, but are not limited to, a layered system architecture, like the one depicted in Figure 1. The lower level of the architecture contains the hardware specific to the RFID infrastructure, which includes, at the minimum, the RFID tags and the readers. Raw RFID data streams generated by the Level-0 devices are transferred to the Edgeware (Level-1), which can be implemented, for instance, using an on-site data server. Depending on the hardware used, the Edgeware is often capable to perform data filtering, cleaning and manipulation. The Edgeware may also instantiate a local database server for buffering and managing the reported data. Finally, the processed RFID data stream is sent to the Middleware (Level-2), whose purpose is to bridge the RFID infrastructure with the upper-level IT applications (Level-3) that rely on its data. The upper level of the architecture in Figure 1 may be further broken down in additional layers (for instance a Service layer and an Applications layer), however in the figure we omit such details as they are not related to the problems we address with our techniques.

We note that unlike the Edgeware, which is typically implemented on-site, the Middleware may be instantiated at a central data processing center. This means that network data movement is required in order to transmit the processed RFID data streams to the Middleware and the applications on top. Furthermore, the Middleware may be responsible for multiple sites equipped with RFID infrastructure. Thus, in order to reduce the network congestion and not to overburden the servers implementing the Middleware, it is desirable that the processed RFID data streams are aggregated as much as possible.

3.2   RFID Data Description

In its simplest form a RFID tag stores a unique identifier called the Electronic Product Code (EPC). When the tag comes in proximity with a reader, the EPC code is read, using an RFID Air Interface protocol, which regulates communication between the reader and the tag. A reader is also equipped with a unique identifier, which in case of immobile readers relates to its location. Finally, an internal clock lets the reader mark the time of the observation. Thus, each time a tag is sensed a triplet of the form

$$(EPC_i, loc_i, t_i)$$

is generated, where EPC_i denotes the code of the tag, loc_i the location (or the id) of the reader and t_i the current time. It is typical in this setup to assume that the time is discretized: a reader reports tags at times t_1, t_2,... where the difference $t_{i+1} - t_i$ is called the *epoch* duration. The reader buffers multiple observations in its local memory and subsequently transmits them to the Edgeware for further processing. This data stream of triplets representing base observations is called the RFID data stream.

The data server at the Edgeware receives the RFID data streams by all readers that it manages in a continuous manner. These readings are filtered based on requirements prescribed by the applications of the upper level. For example EPC codes of locally tagged equipment that are of no interest to the supply-chain monitoring software may be dropped from the stream. A simple form of data reduction is possible at this level my merging occurrences of the same EPC in successive epochs. Let (EPC_i, loc_i, t_i), (EPC_i, loc_i, t_{i+1}),..., (EPC_i, loc_i, t_{i+m}) be part of the stream transmitted by the reader at location loc_i. A straightforward data size reduction is possible if we replace these records with a single quadruple of the form

$$(EPC_i, loc_i, t_{start}, t_{end})$$

with $t_{start}=t_i$ and $t_{end}=t_{i+m}$ indicating the interval containing all observations of tag EPC_i. This is a basic temporal aggregation service that helps reduce the volume of data in the network, when these readings are sent to the Middleware.

Unfortunately, RFID readers routinely drop a significant amount of tag-readings, especially when a large number of tags are concurrently present within the sensing radius of the reader. Moreover, as items are moving withing the facility, the same RFID tag may appear in multiple time intervals in the stream produced by a reader. Both observations complicate management of the RFID data stream and limit the effectiveness of the basic temporal aggregation service.

In our work, we extend the format of the basic tuple generated at the Edgeware to also include a fifth attribute p indicating the percentage of epochs that a tag was observed withing the time interval $[t_{start}, t_{end}]$. A value of p equal to 1 indicates that the tag was spotted during all epochs between t_{start} and t_{end}. This case is equivalent to the format used in basic temporal aggregation. However, a value of p lesser than 1 indicates that only $p \times (t_{end} - t_{start} + 1)$ epochs within the time interval contain observations of the tag. This extension, thus, allows us to represent multiple occurrences of the same tag using fewer intervals. Of course using a single interval to describe all observations of the tag results in large inaccuracy, especially when there are many "holes", i.e. time intervals when the tag was never spotted in the stream. Ideally, given an upper bound on the number of records that can be produced to describe the tag, one would like to find an allocation of intervals that best describe the presence of the tag at the reader.

3.3   Problem Formulation

We can now state our optimization problem formally:

Problem Statement: Given a data stream containing observations of EPC_i at epochs $t_{i_1} < t_{i_2} < \ldots < t_{i_n}$ find the best B-tuple representation $(EPC_i, loc_i, t_{s_1}, t_{e_1}, p_1), \ldots, (EPC_i, loc_i, t_{s_B}, t_{e_B}, p_B)$ where

- $[t_{s_k}, t_{e_k}]$ and $[t_{s_l}, t_{e_l}]$ are non-overlapping intervals ($1 \le k \ne l \le B$),
- $X = \{t_{i_1}, \ldots t_{i_n}\}$, $Y = \{t \in [t_{i_1}, t_{i_n}] | t \notin X\}$, $X \subseteq \cup_k [t_{s_k}, t_{e_k}]$
- $p_k = \frac{|X \cap [t_{s_k}, t_{e_k}]|}{t_{e_k} - t_{s_k} + 1}$

and the cumulative error of the representation

$$\sum_{t \in X} err_x(t) + \sum_{t \in Y} err_y(t)$$

is minimized. Where

$$err_x(t) = \begin{cases} 1 - p_j & , \exists j \text{ such that } t \in [t_{s_j}, t_{e_j}] \\ 1 & , \text{otherwise} \end{cases}$$

and,

$$err_y(t) = \begin{cases} p_j & , \exists j \text{ such that } t \in [t_{s_j}, t_{e_j}] \\ 0 & , \text{otherwise} \end{cases}$$

In this formulation X contains the set of epochs when the tag was spotted by the reader and Y the set of epochs (between its first and last observation) when the tag was not reported. If we use an interval $[t_s, t_e]$ the value of p is determined by the fraction of epochs in X that belong in $[t_s, t_e]$ over the size of the interval. Then the error in estimating the presence of the tag at an epoch t within the interval is (1-p), in case the tag was spotted, and, p in case the tag was not spotted by the reader, respectively. In the formulation of the error function $err_x()$ denotes the false negative error rate when a tag is spotted but we report a value of p less than 1. Similarly, $err_y()$ denotes the false positive error rate when the tag was not spotted by the reader but the epoch in question is inside the interval we report. Thus, our formulation takes into account both false positive and false negative reports of a tag in the computed representation. Given an initial RFID data stream we would like to compute the best representation, using only B tuples, that minimizes the aforementioned error. Obviously, the error is zero when we use as many intervals as the number of epochs in set X. Depending on the distribution of epochs within X, we may be able to derive a much smaller representation with small cumulative error.

In what follows we will discuss algorithms of increased complexity for solving this problem. We first present two observations that help limit the search space in finding the optimal set of tuples.

Lemma 1. *If tuple $(EPC_i, loc_i, t_{s_k}, t_{e_k}, p_k)$ belongs in the optimal B-tuple representation then $t_{s_k}, t_{e_k} \in X$.*

Lemma 2. *If tuple $(EPC_i, loc_i, t_{s_k}, t_{e_k}, p_k)$ belongs in the optimal B-tuple representation then $t_{s_k} - 1, t_{e_k} + 1 \notin X$.*

Lemma 1 states that we should only consider intervals where the starting and ending points both belong to set X. It is easy to see that if one or both end-points do not satisfy this condition, we can always compute a better representation by increasing t_{s_k} (resp. decreasing t_{e_k}) to the nearest epoch when the tag was spotted, as this always reduces the error. Lemma 2 states that when consecutive observations of a tag exist, it is always desirable to package them within the same interval.

4 Temporal RFID Data Aggregation

Recall that given a series of observations of EPC_i at a location, we would like to compute a B-tuple representation of the form $(EPC_i, loc_i, t_{s_1}, t_{e_1}, p_1)$, ..., $(EPC_i, loc_i, t_{s_B}, t_{e_B}, p_B)$. Due to Lemmas 1 and 2, the computation can be performed on the data stream produced after we apply the basic temporal aggregation service. Thus, if we ignore the EPC_i and loc_i values that are constant in this discussion, we are given a set of non-overlapping intervals $[t_{s_1}, t_{e_1}], \ldots, [t_{s_n}, t_{e_n}]$ and would like to replace them with $B << n$ intervals $[t'_{s_k}, t'_{e_k}]$, $1 \leq k \leq B$ such that each input time interval is contained within exactly one of the output intervals. Given the definition of the error we presented in the previous section, we can derive an analytical formula for computing the error associated with a candidate interval $T' = [t'_{s_k}, t'_{e_k}]$ as follows. Let $X(T')$ denote the number of epochs that the tag was reported within T'. Similarly, let $Y(T')$ denote the number of epochs in T', during which the tag was not reported by the reader. Then, the error induced by T' is $(p = \frac{X(T')}{X(T')+Y(T')})$

$$error(T') = \frac{(1-p) \times X(T') + p \times Y(T')}{X(T') + Y(T')} = 2 \times \frac{X(T') \times Y(T')}{(X(T') + Y(T'))^2} \quad (1)$$

4.1   Sub-optimal Algorithms

A straightforward way to obtain a B-tuple representation from the initial n intervals if to first order them by their starting times t_{e_i} and then group them in B batches containing $\lceil \frac{n}{B} \rceil$ input intervals each, except possibly the last. For each batch we generate one interval T' with starting point the starting time of the first input interval in the batch and ending point the ending of the last interval in the batch. The complexity of this simple algorithm is $O(n)$ (linear), assuming that the input n intervals are already ordered by their timestamps. This is a valid assumption, since the basic temporal aggregation process that generates the input intervals operates by first ordering the incoming RFID data stream based on the timestamps of the observations.

The Linear algorithm merges consecutive input intervals, in an error-oblivious manner. A better approximation can be obtained as follows. Given n input intervals, we can consider merging each of the n-1 consecutive pairs in the input. Each candidate pair results in a new interval T' for which we can compute the error using equation 1. A greedy strategy can then be applied that selects the best such interval T' and replaces the corresponding two input intervals with T'. This reduces the number of input intervals by one, to n-1. The same process is then repeated until we are left with B intervals. We call this algorithm the Greedy algorithm. The complexity of the Greedy algorithm is $O((n - B) \times n)$.

4.2   An Optimal Dynamic Programming Algorithm

We now describe a algorithm based on dynamic programing that computes the best B intervals (equivalently best B-tuple representation) that minimize the error of the approximation. Recall that our input consists of n time-intervals that we would like to organize into B non-overlapping intervals, each containing one or more of the original ones. In what follows we would refer to the output intervals that the algorithm considers as *buckets* in order to distinguish them from the input ones. Based on Equation 1, we observe that the error produced by incorporating one or more input intervals within a bucket is independent of the assignments that we have made for other buckets. This observations allows us to state the computation problem using a dynamic programming formulation where the optimal result of a (sub)problem can be obtained by dividing it into two more subproblems and combining the optimal solutions to those subproblems. In particular, let $E(i, k)$ denote the cumulative error of the representation that considers the best way to generate up to k buckets out of the first i input intervals. Obviously $E(i, k)=0$, for $i \leq k$. We can now compute $E(i, k)$ using the following recursion.

$$E(i, k) = \min_{j < i}(E(j, k - 1) + err(j + 1, i)) \tag{2}$$

$err(j+1, i)$ in this formula denotes the error of using a single bucket for merging all input intervals from $j + 1$ up to i. This error is computed by equation 1 by setting $T'=[t_{s_{j+1}}, t_{e_i}]$. Informally, the dynamic programming setup calculates the best way to generate k buckets for the first i intervals by considering the best way to compute k-1 buckets for up to the j-th interval and using a single bucket for intervals $j + 1, j + 2, \ldots, i$. The error of the optimal assignment is calculated by $E(n, B)$ and the optimal bucket configuration arises easily by backtracking the selections made at each step. The running time complexity of the algorithm is $O(n^2 \times B)$.

5   Spatial RFID Data Aggregation

Items tagged with RFIDs are typically moved in groups. For example, packaged products within a pallet are read all together when near an RFID reader. This observation can be exploited by introducing a data representation that *groups* multiple RFID readings within the same temporal record. Given the RFID data stream produced after we apply the temporal aggregation described in the previous Section, we now seek to exploit spatial correlations in its records. In particular, we order the incoming tuples based on their t_{s_k}, t_{e_k} timestamps. Tuples with the same starting and ending timestamps can be encoded using the single common interval by creating a EPC *group-id*. This is a system generated unique code that we will using in order to refer to all EPC_i codes that have been identified during the same time interval. These group-ids are kept in a separate relational table at the Edgeware. This way, they may get possibly reused in order to refer to this subset of codes, when they participate at a larger group in other parts of the data stream.

As an example, consider the records of Table 1 produced after we apply the temporal aggregation process. We can observe that EPC codes I1 and I2 are observed simultaneously at location L1 between T1 and T5. Thus, we can generate a group-id G1 to refer to both products. When these products are later spotted at location $L2$ along with product I4, a new group G2 is generated, which contains both G1 and I4. Table 2 depicts the final set of produced tuples. Table 3 describes the assignment of product codes to group. We note that groups may include other groups (as is the case of G2 and G1). This information needs also be transmitted along with the spatially aggregated tuples in order to decode the groups at the Middleware.

An additional point that we need to clarify when we aggregate multiple records in a group is how to generate a new p-value for the composite record. Recall that the p-values in the temporally aggregated RFID data stream indicate the percentage of epochs that the tag was spotted during the interval indicated in

Table 1. Input RFID Data Stream

EPC	loc	t_s	t_e	p
I1	L1	T1	T5	78
I2	L1	T1	T5	69
I3	L1	T2	T5	90
I1	L2	T12	T22	67
I2	L2	T12	T22	62
I4	L2	T12	T22	66

Table 2. Reduced RFID Data Stream

EPC	loc	t_s	t_e	p
G1	L1	T1	T5	69
I3	L1	T2	T5	90
G2	L2	T12	T22	62

Table 3. Map Table

Group-id	EPC list
G1	I1,I2
G2	G1,I4

the record. When we aggregate several records we have different options in order to produce a new p-value for the group.

- If we seek to reduce the false negative rate of the representation, the p-value of the group should be the minimum of the p-values of all tuples composing the group.
- If we seek to reduce the false positive rate of the representation, we should keep the maximum value of p among the group.
- If we seek to reduce both the false negative and the false positive rates, then we can keep the average value.

In our running example, we used the minimum value, indicating that the application is more concerned about false negative identifications of a tag.

The basic spatial aggregation scheme can be extended in two ways. First, when we compose multiple records, we can set an upper bound on the false positive/negative error rate that we introduce. This may lead to fewer opportunities for spatial aggregation but with reduced error. Another possible extension is to allow an item to partially participate in a group. For example item I3 in Table 1 is spotted with items I1 and I2 in all but the first epoch (T1) at location L1. We could, thus, consider including all three items in a single group for this location.

6   Experiments

In this Section we provide an evaluation of our temporal and spatial aggregation schemes. All algorithms were implemented using Visual Studio 2005. The reported times are on a Intel Core Duo CPU running at 1.83GHz with 1GB of memory. We used as input dataset, the publicly available trace of RFID data obtained during the 2008 Hope Conference in New York, sampled at 30sec intervals. The trace contains 1.9M records of the form (EPC_i, loc_i, t_i). At a first step,

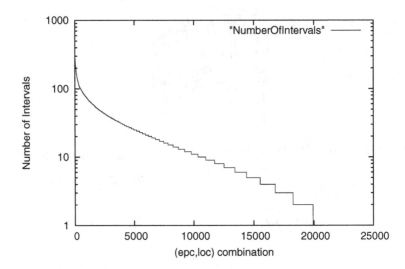

Fig. 2. Number of intervals for each (epc,loc) combination after application of the basic temporal aggregation

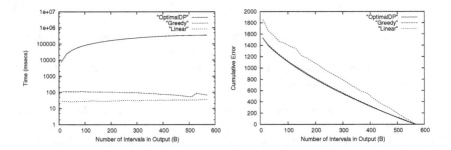

Fig. 3. Execution times   **Fig. 4.** Error of each algorithm

we applied the basic temporal aggregation service that identifies continuous tag observations in order to create tuples of the form (EPC_i, loc_i, t_s, t_e). This process reduced the number of tuples to 423K. The aggregated dataset contained 22,245 unique pairs of (EPC_i, loc_i) values identified at different time intervals.

In Figure 2 we plot the distribution of the number of intervals generated for every pair in the dataset. The higher this number, the more opportunities we have to further reduce the size of the data representation for the particular pair, using our lossy temporal aggregation scheme. We note that, in this dataset, about 10% of the observations are reported in a single interval, which can not be further reduced at the time domain. For the remaining combinations though, we can trade accuracy for compactness in their temporal representation.

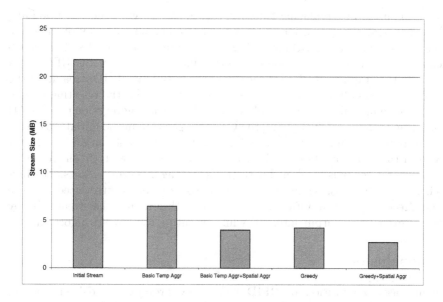

Fig. 5. Combination of Temporal and Spatial Aggregation

In Figure 3 we plot the execution times of the three proposed algorithms: Linear, Greedy and OptimalDP using as input the (tag,loc) combination with the greatest number of intervals (569) from the previous step. We executed the algorithms varying the number of output intervals B requested. Each output interval corresponds to one tuple in the processed RFID data stream. As expected the execution time of Linear is the same independent of the value of B. Greedy works bottom up in constructing the requested number of intervals and its execution time decreases with B. The OptimalDP runs faster when a larger degree of aggregation (smaller value of B is requested). We note that Greedy is up to three orders of magnitude faster than OptimalDP for the same value of B.

In Figure 4 we compare the cumulative error (see Equation 1) computed for the output intervals of each algorithm, when we vary B. As expected, the error of all representations drops when more intervals are used to describe the presence of the tag. We also note, that Greedy computes in most of the cases a representation that is practically the same as optimal. However, this is achieved at a fraction of the time that the dynamic programming algorithm requires, as is evident by Figure 3. We observed this near-optimal behavior of Greedy in all other combinations of tags and locations for this dataset.

We also tested the effect of grouping RFID tags that are moved together over periods of time. We started with the stream resulting from the basic temporal aggregation, which consisted of 423K tuples. Our analysis identified 77K groups of items appearing in identical time intervals. The overall space reduction, which also accounts for the space required for the surrogate group-id descriptions, was 39%. The running time of the spatial aggregation process was 3.3secs. We note that this process did not introduce any error as the original stream can be reverse-engineered by replacing the surrogate group-ids with the corresponding EPCs. The spatial aggregation process can be combined with any of the three temporal aggregation algorithms. In Figure 5 we present the results of an experiment were we tried different methods for reducing the size of the RFID data stream. In the graph we depict (1) the size of the initial raw RFID stream, (2) the size of the stream after the basic temporal aggregation, (3) the resulting stream of the basic temporal aggregation followed by the spatial aggregation process, (4) the resulting stream after applying the Greedy algorithm in the initial dataset in order to reduce the number of tuples for each (epc,loc) combination by a factor of 3:1 (for those combinations with at least three intervals) and, (5) the resulting stream when Greedy is combined with the spatial aggregation process. The first 3 schemes are lossless, while in (4) and (5) some error is introduced because of the Greedy algorithm. Of course, one may further reduce the stream size by choosing even fewer output intervals while executing the Greedy algorithm.

7   Conclusions

The increased adaptation of RFID technology promises to deliver massive datasets. These datasets need to be tamed by reducing their volumes in an application-controllable manner. In this paper we presented several algorithms for temporal and spatial aggregation of RFID data. Our algorithm can reduce the volume of their input data by exploiting correlations at the time and space (location) dimension that characterize the identification of an RFID tag. We provided an experimental study using real RFID traces and demonstrated the effectiveness of our methods.

References

1. Stockman, H.: Communication by Means of Reflected Power. In: IRE (October 1948)
2. Chawathe, S., Krishnamurthy, V., Ramachandran, S., Sarma, S.: Managing RFID Data. In: Proceedings of VLDB, pp. 1189–1195 (2004)
3. Kotidis, Y., Roussopoulos, N.: A Case for Dynamic View Management. ACM Transactions on Database Systems (TODS) 26(4), 388–423 (2001)
4. Jeffery, S., Garofalakis, M., Franklin, M.: Adaptive Cleaning for RFID Data Streams. In: Proceedings of VLDB (2006)
5. Gonzalez, H., Han, J., Li, X., Klabjan, D.: Warehousing and Analyzing Massive RFID Data Sets. In: Proceedings of the 22nd International Conference on Data Engineering (ICDE), p. 83 (2006)
6. Finkenzeller, K., Waddington, R. (eds.): RFID Handbook: Fundamentals and Applications in Contactless Smart Cards and Identification. Wiley, John & Sons, Incorporated, Chichester (2003)
7. Krompass, S., Aulbach, S., Kemper, A.: Data Staging for OLAP- and OLTP-Applications on RFID Data. In: BTW, pp. 542–561 (2007)
8. Park, J., Hong, B., Ban, C.: A Continuous Query Index for Processing Queries on RFID Data Stream. In: 13th IEEE International Conference on Embedded and Real-Time Computing Systems and Applications (RTCSA), pp. 138–145 (2007)

9. Rao, J., Doraiswamy, S., Thakkar, H., Colby, L.S.: A Deferred Cleansing Method for RFID Data Analytics. In: Proceedings of the 32nd international conference on Very large data bases (VLDB), pp. 175–186 (2006)
10. Sarma, S., Weis, S.A., Engels, D.W.: RFID Systems and Security and Privacy Implications. In: Kaliski Jr., B.S., Koç, Ç.K., Paar, C. (eds.) CHES 2002. LNCS, vol. 2523, pp. 454–469. Springer, Heidelberg (2003)
11. Welbourne, E., Koscher, K., Soroush, E., Balazinska, M., Borriello, G.: Longitudinal Study of a Building-wide RFID Ecosystem. In: Mobisys. (2009)
12. Wang, F., Liu, P.: Temporal Management of RFID Data. In: Proceedings of the 31st International Conference on Very Large Data Bases (VLDB), pp. 1128–1139 (2005)
13. Cocci, R., Tran, T., Diao, Y., Shenoy, P.J.: Efficient Data Interpretation and Compression over RFID Streams. In: Proceedings of the 24th International Conference on Data Engineering (ICDE), pp. 1445–1447 (2008)
14. Ioannidis, Y.E.: The History of Histograms (abridged). In: VLDB, pp. 19–30 (2003)
15. Gilbert, A.C., Kotidis, Y., Muthukrishnan, S., Strauss, M.: Optimal and Approximate Computation of Summary Statistics for Range Aggregates. In: PODS (2001)
16. Jagadish, H.V., Koudas, N., Muthukrishnan, S., Poosala, V., Sevcik, K.C., Suel, T.: Optimal Histograms with Quality Guarantees. In: Proceedings of 24th International Conference on Very Large Data Bases (VLDB), pp. 275–286 (1998)
17. Gilbert, A.C., Kotidis, Y., Muthukrishnan, S., Strauss, M.: One-Pass Wavelet Decompositions of Data Streams. IEEE Trans. Knowl. Data Eng. 15(3), 541–554 (2003)
18. Sacharidis, D., Deligiannakis, A., Sellis, T.K.: Hierarchically Compressed Wavelet Synopses. VLDB J. 18(1), 203–231 (2009)
19. Cormode, G., Garofalakis, M.N.: Histograms and Wavelets on Probabilistic Data. In: ICDE, pp. 293–304 (2009)
20. Deligiannakis, A., Kotidis, Y., Roussopoulos, N.: Dissemination of Compressed Historical Information in Sensor Networks. VLDB J. 16(4), 439–461 (2007)
21. Guitton, A., Trigoni, N., Helmer, S.: Fault-Tolerant Compression Algorithms for Delay-Sensitive Sensor Networks with Unreliable Links. In: Nikoletseas, S.E., Chlebus, B.S., Johnson, D.B., Krishnamachari, B. (eds.) DCOSS 2008. LNCS, vol. 5067, pp. 190–203. Springer, Heidelberg (2008)
22. Gonzalez, H., Han, J., Li, X.: Flowcube: Constructuing RFID FlowCubes for Multi-Dimensional Analysis of Commodity Flows. In: Proceedings of the 32nd International Conference on Very Large Data Bases (VLDB), pp. 834–845 (2006)
23. Gray, J., Bosworth, A., Layman, A., Pirahesh, H.: Data Cube: A Relational Aggregation Operator Generalizing Group-By, Cross-Tab, and Sub-Total. In: ICDE, pp. 152–159 (1996)
24. Kotidis, Y.: Extending the Data Warehouse for Service Provisioning Data. Data Knowledge Engineering 59(3), 700–724 (2006)

A Framework for Trajectory Clustering

Elio Masciari

ICAR-CNR
Institute for the High Performance Computing of Italian National Research Council
masciari@icar.cnr.it

Abstract. The increasing availability of huge amounts of "thin" data, i.e. data pertaining to time and positions generated by different sources with a wide variety of technologies (e.g., RFID tags, GPS, GSM networks) leads to large spatio-temporal data collections. Mining such amounts of data is challenging, since the possibility of extracting useful information from this particular type of data is crucial in many application scenarios such as vehicle traffic management, hand-off in cellular networks and supply chain management. In this paper, we address the issue of clustering spatial trajectories. In the context of trajectory data, this problem is even more challenging than in classical transactional relationships, as here we deal with data (trajectories) in which the order of items is relevant. We propose a novel approach based on a suitable regioning strategy and an efficient clustering technique based on edit distance. Experiments performed on real world datasets have confirmed the efficiency and effectiveness of the proposed techniques.

1 Introduction

Trajectories are data logs pertaining time and the position of moving objects (or groups of objects) that could be generated in a wide variety of applications, such as GPS systems [5], supply chain management [16] , vessel classification by satellite images [13].

As an example of such data consider moving vehicles, where both cars and trucks leave a digital trace by the personal or vehicular mobile devices that can be collected via a wireless network infrastructures. Furthermore, mobile phones continuously signalling their locations (cell), are at each moment connected to their GSM network. When the phone sets up a call and moves through the network, there may occur the so called hand-off problem, i.e. the cell where the user is moving through does not have enough bandwidth to accomodate the call. In the same way, GPS-equipped portable devices can record their latitude-longitude position at each moment that they have a location fix, and transmit their trajectories to a collecting server. Moreover, many major retailers ask their provider to exploit RFID technology to tag pallets entering their warehouses and then moving across the supply chain. RFID readers periodically scan the tags appearing in their working area and generate data to report the object being read, the timestamp and the actual location of that object. Finally, a very peculiar type of trajectory is represented by

N. Trigoni, A. Markham, and S. Nawaz (Eds.): GSN 2009, LNCS 5659, pp. 102–111, 2009.

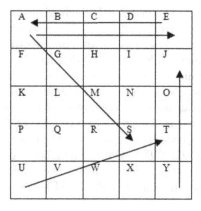

Fig. 1. A set of example trajectories

the stock market. In this case, space can be assumed as a linear sequence of points whose actual values have to be evaluated w.r.t. preceding points in the sequence in order to estimate future fluctuations.

Such a wide spectrum of pervasive and ubiquitous technologies guarantees an increasing availability of large amounts of data pertaining to individual trajectories, having a great localization precision. Therefore, due to the large amount of data generated daily by moving objects, there is a need for analyzing it efficiently in order to extract useful information.

In order to better understand the problem of trajectory clustering which is tackled by the current paper, we provide the following toy example.

Example 1. Consider the trajectories depicted in Fig. 1 and the simple regioning associated to the search space. We can represent the set of trajectories as $T = \{ABCDE, EDCBA, AGMS, UVWST, YTOJ\}$. It is easy to see that while in the transactional case the first two items would be considered the same, in our case they have to be treated separately.

Based on this representation of trajectory data, we propose a novel approach for clustering them based on suitable space partitioning. Indeed, since trajectory data carries information about the actual position and timestamp of a moving object, we can split the search space into regions with suitable granularity and represent them as symbols.

The sequence of regions defines the trajectory traveled by a given object. Note that regioning is a common assumption in trajectory data mining [13,5] and in our case it is even more suitable since our goal is to extract crucial information about trajectories' "shape", i.e. not only the length, but the direction of movement and the eventual turn made during the trajectory.. In this paper we exploit *Principal Component Analysis*(PCA) to effectively identify preferred directions for trajectories.

Once we obtain a proper regioning we can encode the trajectories as string of region symbols since we are interested in identifying the "shape" of the trajectory, while the temporal information relates only to the region traversal order as in the above example. Once the corresponding trajectory strings have been computed we use a suitable string metric based algorithm in order to evaluate the similarity of the trajectories. The most widely known string metric is *Levenshtein* Distance (also known as Edit Distance). This distance metric takes two input strings as arguments, and returns a score equivalent to the number of substitutions and deletions needed in order to transform one input string into another. The rationale for using edit distance is that it is able to capture the main behavior of different trajectories, such as different orders of regions being crossed. In the above example, trajectories $ABCDE$ and $EDCBA$ exhibit the highest dissimilarity as expected.

We performed several tests in order to assess the validity of our proposal and the results so far obtained are convincing, as will be shown in the experimental section.

Outline. In section 2 we formalize the trajectory clustering problem and our regioning strategy exploiting PCA. In section 3 we show the trajectory encoding and the edit distance based clustering strategy. In section 4 we will describe our experimental evaluation. Finally in section 5 we will draw our conclusion.

2   Problem Statement

In this paper we tackle the problem of clustering, from a large body of trajectory data. While for transactional data a tuple is a collection of features, a trajectory is an ordered set (i.e., a sequence) of timestamped items. Trajectory data are usually recorded in variety of different formats, and they can be drawn from a continuous domain. We assume a standard format for each trajectory, with the following definition:

Definition 1 (Trajectory). *Let P and T denote the set of all possible (spatial) positions and all timestamps, respectively. A trajectory is defined as a finite sequence $s_1, \cdots, s_N$, where $N \geq 1$ and each s_i is a pair (p_i, t_i) where $p_i \in P$ and $t_i \in T$.*

Throughout this paper, we assume that P is a set of discrete symbols. For continuous locations, one can partition the space into regions to map the initial locations into discrete symbols. Notice that the method chosen for the assignment of symbols to locations is totally irrelevant to our goal since we are interested in clustering the trajectory as a whole structure. The granularity of the regioning can be decided according to the application requirements. For example, for tracking trucks, the GPS data can be rounded to within 15 meters of precision and so on.

A cluster is a set of regions, and regions belonging to the same cluster are close to each other according to a suitable distance measure. The representative of the cluster is a (possibly imaginary) trajectory that summarizes the main features of the trajectories belonging to the cluster.

2.1   Defining Regions of Interest

The problem of finding a suitable partitioning for both the search space and the actual trajectory is a core problem when dealing with spatial data. Every technique proposed so far, somehow deals with regioning and several have been proposed such as partitioning of the search space in several regions of interest (RoI)[5] and trajectory partitioning [12,8]. In this section, we describe the application of Principal Component Analysis (PCA) in order to obtain a better partitioning. Indeed, PCA finds *preferred* directions for data being analyzed. Thus, we can exploit this information for saving both space and time. Our goal is to find interesting clusters, so it is likely that less frequently crossed regions are not significant.

2.2   Principal Component Analysis

Principal Component Analysis (PCA) finds a linear transformation l which reduces d-dimensional feature vectors to a set of h-dimensional feature vectors (where $h < d$) in such a way that the information is maximally preserved in terms of minimizing the mean squared error. The PCA also allows rolling back to d-dimensions from the h-dimensional feature vectors, with of course the introduction of some reconstruction error.

The h-column vectors define the basis vectors. The first basis vector is in the direction of maximum variance of the given feature vectors. The remaining basis vectors are mutually orthogonal and maximize the remaining variances subject to the orthogonal condition. Each basis vector represents a principal axis. These principal axes are those orthonormal ones onto which the remaining variances are maximum under projection. The orthonormal axes are given by the leading eigenvectors (i.e. those with the largest associated eigenvalues) of the measured covariance matrix. In PCA, the original feature space is characterized by these basis vectors and the number of basis vectors used for characterization is usually less than the dimensionality d of the feature space

Many tools have been implemented for computing PCA such as [19,10] – in our experiments we used a PCA algorithm based on eigenvalue decompositions.

2.3   Exploiting PCA for Regioning

Before defining our simple and effective regioning strategy we must give an intuitive interpretation of the results of PCA analysis on trajectory data. Consider the set of trajectories depicted in Fig. 2.

As shown in Fig. 3 running the PCA algorithm on the above set of trajectories identifies the preferred directions. This information can be exploited using a partition strategy that concentrates on regions along the principal directions. This allows us to focus on the preferred regions when computing clusters.

Indeed, the results of PCA is a set of eigenvalues that states the principal directions among data. Each eigenvalue defines an angle α w.r.t. the principal axes. Depending on the density of data we choose a suitable level of granularity

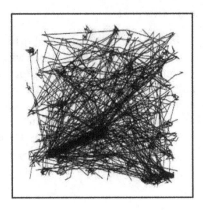

Fig. 2. A set of trajectories

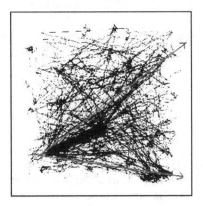

Fig. 3. Principal eigenvalues identified by PCA

that defines the size d of each region (we assume square regions). The size of each region is defined according to the domain being analyzed, for example for cellular nets it could be some meters while for truck movements it could be hundred of meters, while for hurricane control it will be in the order of kilometers. In order to define regions of interest we divide the search space into square regions along the directions set by the eigenvalues so far obtained by PCA and having size d. It will happen that this regioning will also consider regions not parallel to the principal axes, but will not cover the whole search space. Indeed, the "white" regions, i.e. regions not on the principal directions, could be considered as not being interesting regions and thus requiring no further investigation. This strategy is quite effective in practice since we note that infrequently visited regions will not affect clustering results. Thus, pruning the search space will increase the efficiency and effectiveness of the mining step.

2.4   Trajectory Regioning

Once the set of regions $R = \{R_1, R_2, \cdots, R_m\}$ is obtained as explained in the previous sections, we assign to each region a symbol from a given alphabet Σ.

A trajectory Tr_i is a sequence of multidimensional points $Tr_i = p_1, p_2, \cdots, p_n$ where n is the trajectory length. Given a set of trajectories T, we define $T' = encode(T)$ the set of encoded trajectories. An encoded trajectory T'_j is a sequence of regions $T'_j = R_{j,1}, R_{j,2}, \cdots, R_{j,n}$. The encoding is performed by simply substituting each point p_i with the region R_i it belongs to as shown in Fig. 4.

Run PCA(R)
For Each trajectory Tr_i
1:   For each point $p_i \in Tr_i$
2:   find $R_{j,i}$ containing p_i
3:   append $R_{j,i}$ to T'_j

Fig. 4. PCA regioning pseudo code

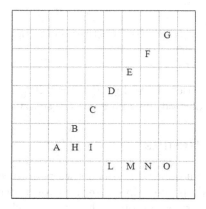

Fig. 5. Regions defined by PCA

Fig. 5 shows a regioning obtained for the set of trajectories from Fig. 3. The encoding phase allows us to obtain a set of trajectories that can be mined using the algorithms that will be explained in next sections.

3   Exploiting Edit Distance for Clustering Trajectories

In order to efficiently identify trajectory clusters, we need to define a measure of similarity between the two sequences. Intuitively, two trajectories are said to have a similar structure if they correspond in the regions crossed and in the order the regions are traversed. Indeed we would like to quantify the similarity between the two trajectories, also emphasizing the differences that are more relevant. For instance, we consider as similar two trajectories that exhibit the same features with different delays since this could be due to a simple traffic problem as will be clear in the next section.

Given two strings their edit distance is defined as the minimum number of operations needed to transform one string into the other. The allowed operations are insertion, deletion, or substitution of a single character, note the the transposition in our case could be seen as the composition of a deletion and an insertion. The string metric so far defined is called *Levenshtein distance*.

Given two encoded trajectories t'_a, t'_b we define $ED(t'_a, t'_b)$ their edit distance. We implemented the algorithm in [14]. In particular we use the following cost assignment:

- copying a region from t'_a over to t'_b (cost 0);
- deleting a region in t'_a (cost 1);
- insert a region in t'_b (cost 1);
- substituting one region for another (cost 1).

Thus $ED(t'_a, t'_b)$ is the minimum among the following:

1. $ED(i-1, j-1) + d(t'_a(i), t'_b(j))$ (substitution or copy);
2. $ED(i-1, j) + 1$ (insert);
3. $ED(i, j-1) + 1$ (delete).

Note that $d(i, j)$ is a function defined as $d(i, j) = 0$ if $t'_a(i) = t'_b(j)$, 1 otherwise, where i and j are symbol positions in a given trajectory.

4   Experimental Results

In this section, we present some experiments we performed to assess the effectiveness of the proposed approach in clustering trajectories. To this purpose, a collection of tests is performed, and in each test some relevant groups of homogeneous trajectories (*trajectory classes*) are considered. The direct result of each test is a similarity matrix representing the degree of similarity for each pair of trajectories in the data set. The evaluation of the results relies on some *a priori* knowledge about the trajectory being used. We performed several experiments on a wide variety of real datasets. We analyzed the following data:

- *School Bus*: a dataset consisting of 145 trajectories of 2 school buses collecting (and delivering) students around Athens metropolitan area in Greece for 108 distinct days[1];
- *Trucks*: a dataset consisting of 276 trajectories of 50 trucks delivering concrete to several construction places around Athens metropolitan area in Greece for 33 distinct days[2];

---

[1] Available at http://www.rtreeportal.org
[2] Available at http://www.rtreeportal.org

- *Animals*: a dataset containing the major habitat variables derived for radio-telemetry studies of elk, mule deer, and cattle at the Starkey Experimental Forest and Range in northeastern Oregon[3].

The similarity matrix enables simple quantitative analyses, aimed at evaluating the resulting intra-class similarities (i.e., the average of the values computed inside each class), and to compare them with the inter-class similarities (i.e., the similarity computed by considering only documents belonging to different classes). To this purpose, values inside the matrix can be aggregated according to the classes of membership of the related elements: given a set of trajectories belonging to n prior classes, a similarity matrix S about these trajectories can be summarized by a $n \times n$ matrix CS, where the generic element $CS(i,j)$ represents the average similarity between class i and class j:

$$CS(i,j) = \begin{cases} \frac{\sum_{x,y \in C_i, x \neq y} ED(x,y)}{|C_i| \times (|C_i| - 1)} & \text{iff } i = j \\ \frac{\sum_{x \in C_i, y \in C_j} ED(x,y)}{|C_i| \times |C_j|} & \text{otherwise} \end{cases}$$

The higher the values on the diagonal of the corresponding CS matrix are w.r.t. those outside the diagonal, the greater the ability of the similarity measure to separate different classes. In the following results we report a similarity matrix for each dataset being considered, as it easy to see it has proven that the technique is quite effective for clustering the datasets being considered.

4.1   School Bus

For this dataset our prior knowledge is the set of trajectories related to the two school buses. We present the results using two classes but we point out that our technique is able to further refine the cluster assignment by identifying the microclusters represented by common subtrajectories. The results are shown in Table 1. It is clear to see from the results that there is a crisp difference between the two clusters.

Table 1. Average similarities for the School Bus dataset

	Bus 1	Bus 2
Bus 1	0.9990	0.7528
Bus 2	0.7528	1

4.2   Trucks

In this case we considered as a class assignment the different trajectories reaching the area where concrete were delivered. In this case there were six main classes as it is shown in Table 2.

---

[3] Available at http://www.fs.fed.us/pnw/starkey/data/tables/index.shtml

Table 2. Average similarities for the Trucks dataset

	Site 1	Site 2	Site 3	Site 4	Site 5	Site 6
Site 1	0.9981	0.7608	0.7610	0.8125	0.8145	0.8275
Site 2	0.7608	0.9909	0.7176	0.7494	0.7150	0.7968
Site 3	0.7610	0.7176	0.9898	0.6054	0.7994	0.7021
Site 4	0.8125	0.7494	0.6054	0.9994	0.7500	0.7515
Site 5	0.8145	0.7150	0.7994	0.7500	0.9970	0.7994
Site 6	0.8275	0.7968	0.7021	0.7515	0.7994	0.9894

4.3 Animals

In this case we considered as a class assignment the different trajectories traversed by elk, mule deer, and cattle. Thus, in this case there were three main classes as shown in Table 3.

Table 3. Average similarities for the Animals dataset

	elk	mule deer	cattle
elk	1.000	0.6639	0.6668
mule deer	0.6639	0.9993	0.7005
cattle	0.6668	0.7005	0.9995

5 Conclusion

In this paper we addressed the problem of detecting clusters in trajectory data. The technique we have proposed is mainly based on the idea of representing a trajectory as string. Thereby, the similarity between two trajectories can be computed by exploiting the edit distance between the associated strings. Experimental results showed the effectiveness of our approach. The current work is subject to further extensions, that we plan to address in future work such as exploiting more information about the regions obtained with PCA and defining a regioning strategy that allows more complex regions. Moreover we would like to investigate a more refined edit distance algorithm in order to better capture the main features of the trajectories being analyzed, i.e. to take into account the Euclidean distances between regions.

References

1. Agrawal, J., Diao, Y., Gyllstrom, D., Immerman, N.: Efficient pattern matching over event streams. In: SIGMOD Conference, pp. 147–160 (2008)
2. Agrawal, R., Faloutsos, C., Swami, A.: Efficient Similarity Search in Sequence Databases. In: Lomet, D.B. (ed.) FODO 1993. LNCS, vol. 730, pp. 69–84. Springer, Heidelberg (1993)

3. Damerau, F.J.: A technique for computer detection and correction of spelling errors. Commun. ACM 7(3), 171–176 (1964)
4. Agrawal, R., et al.: Automatic subspace clustering of high dimensional data for data mining applications. In: SIGMOD (1998)
5. Giannotti, F., Nanni, M., Pinelli, F., Pedreschi, D.: Trajectory pattern mining. In: KDD, pp. 330–339 (2007)
6. Goldin, D., Kanellakis, P.: On similarity queries for time-series data: Constraint specification and implementation. In: Montanari, U., Rossi, F. (eds.) CP 1995. LNCS, vol. 976, pp. 137–153. Springer, Heidelberg (1995)
7. Han, J., Kamber, M.: Data Mining: Concepts and Techniques. Morgan Kaufmann, San Francisco (2000)
8. Jae-Gil, L., Jiawei, H., Kyu-Young, W.: Trajectory clustering: a partition-and-group framework. In: SIGMOD 2007: Proceedings of the 2007 ACM SIGMOD international conference on Management of data, pp. 593–604. ACM, New York (2007)
9. Jeung, H., Yiu, M.L., Zhou, X., Jensen, C.S., Shen, H.T.: Discovery of convoys in trajectory databases. PVLDB 1(1), 1068–1080 (2008)
10. Jolliffe, I.T.: Principal Component Analysis. Springer Series in Statistics (2002)
11. Kéri, G., Kisvölcsey, Á.: On computing the hamming distance. Acta Cybernetica 16(3), 443–449 (2004)
12. Lee, J.-G., Han, J., Li, X.: Trajectory outlier detection: A partition-and-detect framework. In: ICDE, pp. 140–149 (2008)
13. Lee, J.-G., Han, J., Li, X., Gonzalez, H.: *TraClass*: trajectory classification using hierarchical region-based and trajectory-based clustering. PVLDB 1(1), 1081–1094 (2008)
14. Levenshtein: Binary codes capable of correcting deletions, insertions, and reversals. Soviet Physics Doklady 10 (1966)
15. Li, Y., Han, J., Yang, J.: Clustering moving objects. In: KDD, pp. 617–622 (2004)
16. Liu, Y., Chen, L., Pei, J., Chen, Q., Zhao, Y.: Mining frequent trajectory patterns for activity monitoring using radio frequency tag arrays. In: PerCom., pp. 37–46 (2007)
17. Rafiei, D., Mendelzon, A.: Efficient retrieval of similar time series. In: Procs. 5th Int. Conf. of Foundations of Data Organization (FODO 1998) (1998)
18. Sadri, R., Zaniolo, C., Zarkesh, A.M., Adibi, J.: Expressing and optimizing sequence queries in database systems. ACM Trans. Database Syst. 29(2), 282–318 (2004)
19. Sharma, A., Paliwal, K.K.: Fast principal component analysis using fixed-point algorithm. Pattern Recognition Letters 28(10), 1151–1155 (2007)
20. Yang, J., Hu, M.: Trajpattern: Mining sequential patterns from imprecise trajectories of mobile objects. In: Ioannidis, Y., Scholl, M.H., Schmidt, J.W., Matthes, F., Hatzopoulos, M., Böhm, K., Kemper, A., Grust, T., Böhm, C. (eds.) EDBT 2006. LNCS, vol. 3896, pp. 664–681. Springer, Heidelberg (2006)

Preliminaries for Topological Change Detection Using Sensor Networks

Jixiang Jiang and Michael Worboys

NCGIA, University of Maine, ME, 04468, USA
{jixiang.jiang,worboys}@spatial.maine.edu

Abstract. Topological changes to regions, such as merging/splitting and hole formation/elimination, are significant events in their evolution. Information about such salient changes is useful in many applications. The research reported in this paper provides theoretical foundations for such topological change detection in sensor networks. A local tree model is proposed in the spatial domain, based on which a set of types of topological changes is specified. We also present a sensor network framework which captures the necessary information required by the tree model. Both the local tree model and the sensor network framework form the foundations for detection approaches that allow sensor networks to report topological changes.

1 Introduction and Background

Wireless sensor network technology provides real-time information about the environment, and this will play an important role in the monitoring of geographic phenomena. Most sensing applications up to now have focused on capturing, processing and reporting geographical information in the form of spatial-temporal data. However, topological changes to regions, such as merging/splitting and hole formation/elimination, are often the significant events, and in many applications it is useful to have information about such topological changes. For example, in the case of wildfire, fire fighters might be interested if the fire zone regions *split* and become disconnected, so that they can reorganize the team accordingly. They might also be interested in *merging* fires, as it sometimes slows down the burn when the fires are burning over each other. This paper focuses on addressing the basic issues related to the application of sensor networks to the detection of topological changes.

Topological features are important aspects of spatial data. Their representations are the focus of much research work in spatial data modeling. Topological features allow us to classify spatial changes. Egenhofer and Al-Taha [1] analyze and classify the spatial changes involving two regions based on their topological relations before and after the change, and the result is recorded using the conceptual neighborhood graph. In previous work [2], we have proposed a model that represents the dynamic topology of an areal object (a collection of region components, possibly with holes and islands), based on which different types of topological changes are specified.

N. Trigoni, A. Markham, and S. Nawaz (Eds.): GSN 2009, LNCS 5659, pp. 112–121, 2009.
© Springer-Verlag Berlin Heidelberg 2009

A straightforward application of sensor networks is to the monitoring of geographical phenomena [3]. Previous research either focuses on proposing energy-efficient approaches to transmitting entire sensing data back to base stations [4,5], or focuses on providing important spatial properties of the phenomena. For example, the snake model proposed by Jin and Nittel [6] is able to derive the area and centroid of a deformable 2D object over time. Recently, there is an increasing interest in considering topological information when processing sensing data. Gandhi, Hershberger and Suri [7] emphasize the topology of the isolines in a scalar field and propose an approach that approximate a family of isolines by a collection of topology-preserving polygons. Sarkar *et al.* [8] present a distributed algorithm for the construction of a contour tree to represent the topological structure of contours in a scalar field, based on which isoline queries can be enabled. Worboys and Duckham [9] provide a computational model for sensor networks to detect global high-level topological changes based on low-level 'snapshot' of spatiotemporal data.

Current research has made initial attempts to topological change detection using wireless sensor networks. Farah and Cheng *et al.* [10] provided initial attempts to detect topological changes in responsive sensor networks by an event-driven approach. Sadeq [11] proposed the idea of detecting topological changes by maintaining the boundary state of areal objects. In [12], we presented the topological change detection approach based on local aggregation.

2   Local Tree Model

This work is concerned with evolving areal objects in the plane. An areal object is a collection of region components, possibly with holes or islands. At a particular time, an areal object can be considered as a set R of points in the spatial domain, and the evolution of an areal object can be considered as a temporal sequence of snapshots of the areal object. Each pair of consecutive snapshots describes a change, called a *basic transition*. Topological changes can be observed during a basic transition.

Let R_1 and R_2 be a pair of areal objects that define a basic transition, where R_1 and R_2 are the start and end snapshots, respectively. Any location p in the spatial domain must have one of the following four states: (I) $p \notin R_1$ and $p \in R_2$. (II) $p \in R_1$ and $p \notin R_2$. (III) $p \in R_1$ and $p \in R_2$, or (IV) $p \notin R_1$ and $p \notin R_2$. Based on the states of each location, the whole spatial domain can be partitioned into several components. Each component is a maximal topologically connected set of locations in the same state.

The locations in state I or state II form the components that are either added to or removed from the areal object during the basic transition. We call such components *transition regions*. For simplicity, we assume each basic transition has only one transition region that is topologically equivalent to a disk; i.e., the transition region is a single piece without any holes. The locations in state III or IV form the components that do not change during the basic transition. The structure of such components is important to determine the type of the basic

Fig. 1. A basic transition

transition. However, not all of the components are crucial in the determination of the type of topological change. Only the components that are adjacent to the transition region are necessary. These components are referred to as *C-components*.

As an example, Fig. 1(a) shows a basic transition of an areal object, based on which the spatial domain is partitioned into five components *A-E*, as shown in Fig. 1(b). Among these components, component C is the transition component, and components A, B and D are C-components.

To specify topological changes, we are most interested in the adjacency relations and the surrounded-by relations between the C-components. These are defined as follows:

Definition 1. *Let X_1 and X_2 be a pair of components in a basic transition.*

1. *X_1 is said to be* adjacent *to X_2 if the boundary of X_1 intersects the boundary of X_2.*
2. *X_1 is said to be* surrounded by *X_2 if any path that connects a point in the closure of X_1 to a point at infinity intersects the closure of $X_2 \backslash X_1$. X_1 is said to* surround *X_2, if X_2 is surrounded by X_1.*

The structure of the C-components in a basic transition have the properties stated as follows, and section 4 provides detailed proofs of these properties.

1. There is exactly one C-component X which surrounds all the other C-components. X is referred to as the *background C-component* of the basic transition.
2. The topological structure of the C-components in a basic transition can be represented by a rooted tree. A vertex of the tree represents a C-component, and an edge of the tree connects a pair of vertices representing adjacent C-components. The root of the tree represents the background C-component.

As different rooted trees can be explored in a systematic way, we are able to generate the possible topological structures between the C-components of a basic transition. Fig. 2 lists all the rooted trees with less than 4 vertices, and examples of structures represented by the rooted trees are also provided. In the figure, the transition region is indicated by shaded area, and the vertices of the representation tree are placed inside the C-components they represent.

The classification of a basic transition is based on the following three factors: (1) The topological structure of its C-components, (2) the state of the locations

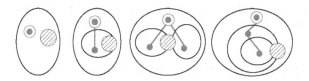

Fig. 2. Tree representations for different configurations of C-components

Local topological structure	State of transition region (T)		State of background C-component (B)	
	T = I B = IV	T = II B = IV	T = I B = III	T = II B = III
◉	Region appear (A)	Region disappear (B)	Hole disappear (I)	Hole appear (J)
⦙	Topology-preserving changes (C, D, K, L)			
⋀	Regions merge (E)	Region split (F)	Hole split (M)	Hole merge (N)
⦙	Region self-merge (G)	Region self-split (H)	Hole self-split (O)	Hole self-merge (P)

Fig. 3. Classification of basic transitions

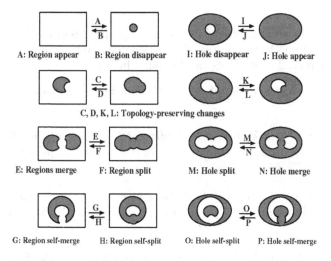

A: Region appear   B: Region disappear   I: Hole disappear   J: Hole appear

C, D, K, L: Topology-preserving changes

E: Regions merge   F: Region split   M: Hole split   N: Hole merge

G: Region self-merge   H: Region self-split   O: Hole self-split   P: Hole self-merge

Fig. 4. Examples of specific types of topological changes

in its transition region, and (3) the state of the locations in its background C-component. The classification yields different types of topological changes incurred by a basic transition. Fig. 3 shows the classification results, in which the 13 types of specific topological changes are distinguished, and Fig. 4 provides an example of a basic transition for each type of topological change. These types of topological changes will be used in the sensor report to describe the observed basic transition.

3   Completeness of Local Tree Model

In this section, we prove that the structure of C-components in any basic transition can be represented by a rooted tree, and therefore the local tree model provides a complete coverage over all basic transitions.

The following lemmas and theorems use the notion of 'partially surrounded-by'. Let X_1 and X_2 be a pair of distinctive C-components in a basic transition, and let T be the transition region. X_1 is defined to be *partially surrounded by* X_2 (or X_2 *partially surrounds* X_1) if X_1 is surrounded by $T \cup X_2$. Fig. 5 shows an example in which C-component X_1 is partially surrounded by C-component X_2.

Fig. 5. Example of partially-surrounded-by relation

We first present several lemmas to show the properties of the C-components in a basic transition. As the proofs of these lemmas are long, we do not include them in this paper. These proofs can be found in [13].

Lemma 1. *The adjacency graph of the C-components in a basic transition is connected.*

Lemma 2. *Let X_1 and X_2 be a pair of adjacent C-components in a basic transition. One of the following statements must be true: either X_1 partially surrounds X_2, or X_2 partially surrounds X_1.*

Lemma 3. *Let X_1, X_2, and Y be distinct components in the spatial domain such that $X_1 \cup Y$ surrounds X_2 and $X_2 \cup Y$ surrounds X_1. Y must surround both X_1 and X_2.*

Lemma 4. *Let X_1 and X_2 be a pair of C-components. The following statements cannot both be true: (1) X_1 is partially surrounded by X_2. (2) X_2 is partially surrounded by X_1.*

Lemma 5. *Let X_1, X_2 and Y be distinct C-components. The following statements cannot both be true: (1) X_1 is adjacent to Y and partially surrounds Y. (2) X_2 is adjacent to Y and partially surrounds Y.*

Theorems 1 and 2 present the major results of this section.

Theorem 1. *The adjacency graph of the C-components in a basic transition is a tree.*

Proof. This can be proved by showing that the adjacency graph of the C-components in a basic transition is both connected and cycle-free.

By Lemma 1, the adjacency graph is connected. We prove it is cycle-free by contradiction. Suppose we are able to find a cycle in the adjacency graph, which must contain more than 2 vertices. By Lemma 2, the partially-surrounded-by relation must hold between the C-components represented by any pair of adjacent vertices in the cycle, and there are two possible cases to consider:

Case 1: there are three consecutive vertices in the cycle representing C-components X_{i-1}, X_i, and X_{i+1}, respectively, such that both X_{i-1} and X_{i+1} are adjacent to X_i and partially surround X_i. This contradicts the conclusion of Lemma 5.

Case 2: for $\forall i$, $i = 1, 2, ..., k - 1$ (where k is the number of vertices in the cycle), X_i partially surrounds X_{i+1}, and X_k partially surrounds X_1. It follows that X_1 partially surrounds itself. Hence by definition, the transition region surrounds X_1. This contradicts the assumption that the transition region is simply connected.

Both cases lead to contradictions, so the adjacency graph of the C-components is both connected and cycle-free, and therefore must be a tree.   □

Theorem 2. *Among all the C-components of a basic transition, there must be exactly one C-component that surrounds all the other C-components.*

Proof. It is straightforward to prove this theorem, if there is one or two C-components in the basic transition. Consider the basic transitions which have more than two C-components.

First, we show that there must be at least one C-component B that partially surrounds all the other C-components. If not, we are able to find two C-components X_n and X_m such that neither of them is partially surrounded by any other C-components. By theorem 1, the adjacency graph of these C-components is a tree. By Lemma 2, in the path of the adjacency tree that connects the vertices representing X_n and X_m, there must be three consecutive vertices that represent C-components X_{i-1}, X_i and X_{i+1}, such that both X_{i-1} and X_{i+1} are adjacent to X_i and partially surround X_i. This contradicts the result of Lemma 5. Therefore, there must be at least one C-component B that partially surrounds all the other C-components.

In addition, by Lemma 4, there is at most one C-component B that partially surrounds all the other C-components. In all, there is exactly one C-component B that partially surrounds all the other C-components.

Finally, we show that B surrounds all the other C-components. Let T be the transition region, and U be the union of all the C-components except for B. As B partially surrounds all the other C-components, it follows that $B \cup T$ surrounds U. In addition, by the definition of C-components, $B \cup U$ surrounds T. By Lemma 3, we have B surrounds U, and therefore B surrounds all the other C-components.   □

Theorems 1 and 2 show that the structure of C-components in any basic transition can be represented by a rooted tree, in which the root represents the unique background C-component.

4   Sensor Network Configuration

We are using sensor networks to track and report topological changes. This section provides the basic assumptions we have made about the sensor networks, as well as the definitions of basic elements in sensor networks that are approximate representations of the components and relations required by the local tree model.

We assume that a node located near the boundary of the sensing area is selected to be the *reference node*, which is assumed to be located outside the scope of the observing phenomena. The sensor nodes in the sensing area induce a Voronoi diagram, and each sensor node n is associated with a Voronoi cell consisting of all the locations that are closer to n than to any other sensor node.

The sensor nodes take measurements at a sequence of sampling rounds t_0, t_1, ..., t_n. We assume that the reading of a sensor node at any of the sampling rounds is either 0 or 1. Our interpretation is that the reading is 1 if the sensor node is in an area of high intensity (reading above a threshold), otherwise it is 0. A change is captured by sensor readings at a pair of consecutive sampling rounds, and the type of a transition is determined by comparing the readings. The comparison first defines four states of nodes at sampling round t_i.

Definition 2. *Let $r(n,t) \in \{0,1\}$ denote the reading of a node n at a time t. The state of n at a sampling round $t_i (1 \leq i \leq k)$ is defined to be a pair $h = (r(n, t_{i-1}), r(n, t_i))$, such that $h \in \{(0,1), (0,0), (1,0), (1,1)\}$.*

The states of the sensor nodes together with the sensor connectivity yield the following concepts that approximate the fundamental properties required by the local tree model.

Definition 3. *Let N be a set of sensor nodes, N is said to be a* homogeneous sensor component *if the nodes in N are in the same state and induce a connected component in the communication graph. Moreover, N is defined to be a* maximal homogeneous sensor component, *if it is impossible to find a node n in the sensing area such that (1) $n \notin N$, and (2) $N \cup \{n\}$ is a homogeneous sensor component.*

Definition 4. *Let N_1 and N_2 ($N_1 \cap N_2 = \emptyset$) be a pair of homogeneous sensor components.*

1. *N_1 is said to be* adjacent *to N_2 if there are nodes $n_1 \in N_1$ and $n_2 \in N_2$ such that n_1 and n_2 are direct neighbors in the communication graph. Otherwise, N_1 and N_2 are said to be* separated.
2. *N_1 is said to be* surrounded *by N_2, if any path in the communication graph that starts from the reference node and contains a node of N_1 must contain a node of N_2. N_1 is said to* surround *N_2, if N_2 is surrounded by N_1.*

Definition 5. *Let N be a maximal homogeneous sensor component.*

1. *N is defined to be a* transition sensor component, *if N consists of only nodes either in state* $(0, 1)$ *or in state* $(1, 0)$.
2. *N is defined to be a* sensor C-component, *if both of the following conditions are satisfied: (1) N consists of nodes either in state* $(0, 0)$ *or in state* $(1, 1)$, *and (2) N is adjacent to transition sensor component.*
3. *N is defined to be a* background sensor C-component, *if it is a sensor C-component and it surrounds all the other sensor C-components.*

Table 1. Elements in spatial domain and their approximations in sensor networks

In spatial domain	In sensor networks
A C-component in state III	A sensor C-component in state $(1, 1)$
A C-component in state IV	A sensor C-component in state $(0, 0)$
The transition region	The transition sensor component
The background C-component	The background sensor C-component
Adjacency relations between C-components	Adjacency relations between sensor C-components
Surrounded-by relations between C-components	Surrounded-by relations between sensor C-components

Table 1 shows the correspondences between the components and relations we defined in spatial domain and their representations in sensor networks. Based on the correspondences, all the concepts defined in the spatial domain can be represented and computed in terms of states of sensors and connectivity between them.

Ideally, the elements defined in sensor networks represent the properties of the areal objects in the spatial domain. The nodes located in a component of the spatial domain form exactly one maximal homogeneous sensor component. Components in the spatial domain are adjacent if and only if they are represented by adjacent maximal homogeneous sensor components. Components in the spatial domain surround each other if and only if they are represented by maximal sensor components that surround each other. However, such a perfect matching may not always exist. Inconsistency may be caused by low node density and improper setting of communication ranges. Here are some of the examples.

First, if the density of the nodes is low, a component in the spatial domain that is small enough may not contain any sensor node, and therefore is not represented by any sensor component, as shown in Fig. 6(a).

Second, if the communication range of the sensor nodes is not large enough, inconsistencies may occur. For example, a pair of adjacent components in the spatial domain are represented by a pair of separated maximal homogeneous sensor components, as shown in Fig. 6(b).

Finally, if the communication range of the sensor nodes is not small enough, inconsistencies may occur. For example, a pair of components that are not adjacent in the spatial domain is represented by a pair of adjacent maximal homogeneous sensor components, as shown in Fig. 6(c).

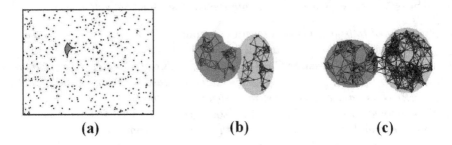

(a)   (b)   (c)

Fig. 6. Configurations that lead to errors in reports

In order to avoid inconsistency, we stipulate that the sensor node deployment satisfies the following constraints: **(1) Density constraint,** sensor nodes are deployed densely enough so that a sensor node measurement accurately reflects all locations in its Voronoi cell. **(2) Communication constraint,** each sensor node communicates exactly with the nodes in its adjacent Voronoi cells.

Sensor networks that conform to both constraints are able to provide correct sensing reports on the topological changes based on the framework we provided. Failure to conform to both constraints does not disable the whole detection approach, but it might result in errors in the reports of some sensing rounds. For example, sensor network might report that splitting of a wild fire is observed in a sensing round, but in reality the fire does not split.

5   Conclusion and Future Work

This paper provides the computational foundations for topological change detection in sensor networks. Based on the local tree model, basic transitions are classified into a set of classes, and each class specifies a type of topological change. We also analyze the corresponding elements required by the local tree model in the sensor network configuration. Based on these foundations, we have developed distributed algorithms for topological change detection using sensor networks, detailed in [13].

Future work includes the following areas: (1) extensions of current research in order to handle non-incremental transitions, in which more than one transition region exists in a pair of consecutive snapshots, and each transition region may have holes. (2) design of algorithms for topological change detection under imperfect information, which can be introduced in either by uncertainty in sensor readings or by improper configuration of sensor networks.

Acknowledgments

This material is based upon work supported by the National Science Foundation under grant numbers IIS-0429644 and IIS-0534429. Michael Worboys' work is also supported by the National Science Foundation under NSF grant number DGE-0504494.

References

1. Egenhofer, M., Al-Taha, K.: Reasoning about gradual changes of topological relationships. In: Frank, A.U., Formentini, U., Campari, I. (eds.) GIS 1992. LNCS, vol. 639, pp. 196–219. Springer, Heidelberg (1992)
2. Jiang, J., Worboys, M.: Event-based topology for dynamic planar areal objects. IJGIS 23(1), 33–60 (2009)
3. Nittel, S., Stefanidis, A., Cruz, I., Egenhofer, M., Goldin, D., Howard, A., Labrinidis, A., Madden, S., Voisard, A., Worboys, M.: Report from the First Workshop on Geo Sensor Networks. ACM SIGMOD Record 33(1), 141–144 (2004)
4. Sharaf, M., Beaver, J., Labrinidis, A., Chrysanthis, P.: TiNA: A scheme for temporal coherency-aware in-network aggregation. In: Proceedings of the 5th International ACM Workshop on Data Engineering for Wireless and Mobile Access, pp. 69–76 (2003)
5. Silberstein, A., Braynard, R., Yang, J.: Constraint chaining: On energy-efficient continuous monitoring in sensor networks. In: Proceedings of SIGMOD 2006, pp. 157–168 (2006)
6. Jin, G., Nittel, S.: Tracking deformable 2D objects in wireless sensor networks. In: Proceedings of ACM-GIS 2008, pp. 491–494 (2008)
7. Gandhi, S., Hershberger, J., Suri, S.: Approximate isocontours and spatial summaries for sensor networks. In: Proceedings of IPSN 2007, Cambridge, MA, USA, pp. 400–409 (2007)
8. Sarkar, R., Zhu, X., Gao, J., Guibas, L., Mitchell, J.: Iso-contour queries and gradient descent with guaranteed delivery in sensor networks. In: Proceedings of INFOCOM 2008, Phoenix, AZ, USA (2008)
9. Worboys, M., Duckham, M.: Monitoring qualitative spatiotemporal change for geosensor networks. IJGIS 20(10), 1087–1108 (2006)
10. Farah, C., Zhong, C., Worboys, M., Nittel, S.: Detecting topological change using wireless sensor networks. In: Cova, T.J., Miller, H.J., Beard, K., Frank, A.U., Goodchild, M.F. (eds.) GIScience 2008. LNCS, vol. 5266, pp. 55–69. Springer, Heidelberg (2008)
11. Sadeq, M.: Distributed Detection of Spatiotemporal Change of Regions in Wireless Sensor Networks Using Boundary State. In: Proceedings of DG/SUM 2007 (2007)
12. Jiang, J., Worboys, M.: Detecting basic topological changes in sensor networks by local aggregation. In: Proceedings of ACM-GIS 2008, pp. 13–22 (2008)
13. Jiang, J.: Specifying and detecting topological changes to areal objects. PhD thesis, The University of Maine (2009)

On the Feasibility of Early Detection of Environmental Events through Wireless Sensor Networks and the Use of 802.15.4 and GPRS

Jorge Santos, Rodrigo M. Santos, and Javier D. Orozco

Instituto de Investigaciones en Ingeniería Eléctrica
Universidad Nacional del Sur - CONICET
Av. Alem 1253 - (8000) Baha Blanca - Buenos Aires - Argentina
iesantos@criba.edu.ar, ierms@criba.edu.ar, jorozco@uns.edu.ar

Abstract. In this paper, a general model of an Early Detection of Environmental Events System is presented. It follows established technologies and it is based on wireless sensor network nodes connected to a central computer using IEEE 802.15.4 and the General Radio Packet Service. In order to test its feasibility, the model is applied to the early detection of forest fires. The contribution of this paper is to show that although the system is theoretically apt to work well in small grid squares, its deployment over the totality of large areas is unfeasible due to the sheer number of sensors necessary, and their present costs and deployment difficulties. Instead, a second best alternative, giving not an early warning of an impending fire but an early notice that the fire is in its beginning phase, is shown to be feasible, from both technical and economic points of view.

Keywords: Early warning, wireless sensor network, environmental events.

1 Introduction

Very often started by accident, environmental events can cause considerable damage to the environment itself as well as to property and/or people. Usually, it is hard to predict when, where and how this kind of incidents will begin. However, in many cases it is possible to devise a system that, based on the retrieval of a well chosen set of data, anticipates the beginning of this kind of events or, as a second-best alternative, detects the event immediately after it actually took place and is still in its starting phase. Such a system will be called Early Detection of Environmental Events System (EDEES).

An EDEES can be used for many purposes, for instance to detect forest fires, river floods, volcanic activity, seismic activity, etc. They are a paradigmatic real-time application: since the instant of an anomaly detection in the field until taking appropriate countermeasures, the delay must be bounded for the countermeasures to be effective.

Wireless Sensor Networks (WSNs) are increasingly proposed to perform surveillance on difficult access places or remote locations, where the presence of human

N. Trigoni, A. Markham, and S. Nawaz (Eds.): GSN 2009, LNCS 5659, pp. 122–130, 2009.

operators is impractical. For this purpose, WSNs rely on small System-On-a-Chip devices called *motes*; by combining sensor, calculation and transmission capabilities, they can sense some variables and transmit their values, raw or processed, to a central computer where decisions may be taken according to the available data. WSNs therefore seem to be a good choice to implement an EDEES.

In this paper, a general EDEES model and its application to two alternatives of forest fire detection systems are presented. Both alternatives are based on the use of sensors of weather and field variables, the IEEE 802.15.4 and the General Packet Radio Service as communication protocols, and a central computer to process the acquired data. A combined time-driven/query-driven approach is used there to implement the real-time surveillance.

The rest of the paper is organized as follows. In Section 2 related work is discussed. In Section 3, the general system model is presented. In Section 4, its application to two alternatives of EDEES dedicated to forest fire detection are proposed and their feasibility analysed. Finally, in Section 5, conclusions are drawn.

2 Related Work

In the summer of 2007, an extensive bibliography was published by the Autonomous Network Research Group of the University of California, under the direction of Dr. B. Krishnamachari. It contains over 1000 references, mostly of the previous five years. In what follows, only a few are discussed, those presenting calculation methods used in this paper or proposing forest fire early warning systems, are commented.

In [1], a WSN is used to monitor forest fires. In the paper, the authors describe an algorithm based on the computation of the Canadian Fire Weather Index (FWI). There is no response time analysis. The approach is repeated in [2] although the authors use a different index to measure the probability of fire.

In [3], a forest fire monitoring WSN is also proposed. Each node of the network can produce a regular report periodically, an emergency report upon a special event and finally may answer to the query of the master. Once the information has been collected, a neural network processes the retrieved data and produces an output indicating the probability of a developing fire in the area.

In [4,5,6], IEEE 802.15.4 is studied in detail. The analysis gives a full understanding of the behaviour of the deterministic mechanism, to operate in real-time, with regard to delay and throughput metrics. The network scheduling is analysed under different modes of operation. The calculation methods are used in this paper.

In [7], exhaustive simulation-based performance evaluations of the Medium Access Control sublayer of the standard are presented.

In [8], an analysis of the cost/benefit trade-off in sensor networks is presented. The authors discuss the utility of using this kind of technology and its efficiency versus more traditional ones. Special consideration is given to large area coverage where the deployment of sensors is difficult or expensive.

3   System Model

In general, EDEESs are based on the acquisition of data in situ. Data, raw or with some sort of elaboration, must be transmitted through communication networks to a central computer (CC), which processes them and, if necessary, raises the alarm to prompt countermeasures. Clear indications about the geographical location and type of detected event must be given. In what follows, the type of variables to be sensed, the networks to transmit their values, and their processing are described.

3.1   Indices

The sensors measure weather and field variables, for instance air temperature, wind direction, wind intensity, humidity, UV intensity, barometric pressure, soil moisture, ground temperature, precipitations (rain, dew) etc. The values of these variables, captured by the WSN, are fused and an index is built. The probability of an incident, based on the value of the index, is then estimated by the central computer implementing a real-time policy. Obviously, different types of events (fires, floods, earthquakes, volcanoes, etc.) require different types of indices. Moreover, for the same type of events, indices of different complexity can be devised. As a rule of thumb it can be said that more complexity in the index is roughly associated to an earlier warning of the impending event. However, more complex indices generally increment the cost of the sensors and the load on calculations and communications, leading to a shorter life of the batteries, a relevant issue because the lifespan of a mote is the life of its battery. The trade-off point between complexity of the index and life of the battery must be set by the designer.

3.2   802.15.4

At the base of the EDEES is a WSN whose nodes are implemented by motes. The IEEE 802.15.4 standard has been adopted by many manufacturers of motes [9]. It provides wireless connectivity, with very low complexity, cost and power, among low data-rate cheap devices. Therefore it is used as the lower communication protocol of the EDEES.

IEEE 802.15.4 covers only the two lower layers of the Open System Interconnection model. The higher layers are covered by ZigBee, a protocol developed by a consortium of vendors to enhance the IEEE Standard. It will be used in what follows.

3.2.1   Topologies

The standard proposes three different topologies for the network: star, mesh and cluster-tree. In the first one, there is a central node at the hub of the star. The rest of the nodes communicate only with the central one. In the mesh topology, a direct link can be established between any two nodes that are within range. Finally, the cluster-tree combines the previous two. The use of each one depends

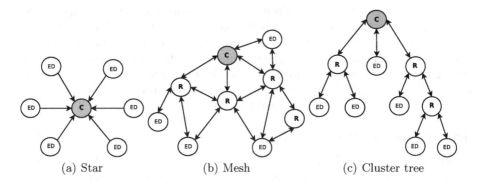

(a) Star	(b) Mesh	(c) Cluster tree

Fig. 1. 802.15.4/ZigBee Topologies

on the application and the physical constraints of the place where the WSN will be deployed. It must be noted that the range of a mote is no more than 30 meters, limiting in this way the distance between adjacent nodes with the possibility of a direct link. This limits the star topology to a circular area with a diameter of at most 60 m.

The mesh topology permits a wider coverage since nodes can communicate through other nodes; by doing so a multi-hop path is established reaching very long distances. The information can be spread among all nodes and a full distributed system may be implemented on top of the WSN. However, the mesh topology only admits a non deterministic access mode to the transmission medium and therefore it cannot be used in a real-time application. On top of that, the number of nodes involved in an environmental monitoring application is usually in the order of thousands, making the routing tables in the nodes very large and so making difficult the administration of the network.

Finally, the cluster-tree topology combines the predictability of the star topology and the wider range of the mesh one. Three different kind of nodes are used: Coordinators, Routers and End Devices. A Coordinator (C), at the root of the tree, is in charge of choosing a Personal Area Network identifier, not used by any other neighbour. Routers (R) communicate with the Coordinator and negotiate with it the address assignment, communication parameters, and the possibilities of extending the network. The sensors are the End Devices (ED) and, therefore, the leaves of the tree.

In order to provide coverage to a wide area, some other technology like radio, GPRS, or even cell-phone links, have to be incorporated. Therefore, the C-nodes in Figure 1 are also part of a higher order network.

3.2.2 Medium Access Control

802.15.4 has two layers, the Physical and the Medium Access Control (MAC) in which the rules to access the transmission medium are specified. There are two operational modes: *the non beacon-enabled* and *the beacon-enabled*. The second one provides timeliness guarantees and is, therefore, the one used in this system. In this mode, communications are synchronized by a beacon. Each Beacon

Interval has an active part in which data is transmitted within a Superframe structure and an inactive part in which nodes may sleep, so extending the life of their batteries. The Superframe has three different parts. In the third one, a Contention Free Period access to the medium is granted in the form of Guaranteed Time Slots. The use of this mechanism bounds delays and makes possible the granting of real-time guarantees. An excellent description of the Guaranteed Time Slots management can be found in [6].

3.3   The Higher Order Network

Because of its wide availability, General Packet Radio Service (GPRS) is a good candidate for the higher order network. It is a packet-based communication service for mobile devices that allows data to be sent and received across a mobile telephone network. It is a technology halfway between the second and third generation of cell-phones. Its theoretical bandwidth is 171.2 kbps but it usually operates at a much lower rate, 20-50kbps, on top of older GSM networks; therefore, its coverage is usually large.

The values of the variables may be preprocessed in the cluster and sent to the central computer through the GPRS node with 16 bit precision. The GPRS packets have a maximum of 1600 bytes, and all the information can be sent to the CC in a single packet. In [10] a delay and throughput analysis is presented for GPRS, based on the packet size and the channel interference. In our case, using the methodology proposed in [10], there is a worst delay of 0.5 s for a packet size of 400 bytes, long enough to transmit all the necessary information from the C-node to the CC.

3.4   How the System Operates

The system is capable of operating in two modes: event-driven and query-driven. In the first mode, the event that triggers the transmission from the cluster to the CC is the detection of an index value pointing to the initiation of an event. A typical delay from an ED-node to a C-node up the levels of the tree plus the computation of the index is 10 s; transmitting through the GPRS may take up to 0.5 s.

The query-driven mode is used to verify that the whole system is operative. The CC queries all the clusters in a round-robin fashion. If an answer is not given or it is incorrect, the query process may be repeated up to three times. If the fault persists, it may be assumed that it is permanent and must be attended to. Faults at the nodes of the mesh may be caused, for instance, by wind action, falling branches, wild animals, etc.

Since the motes consume more energy while listening or transmitting, the implementation of a low duty query cycle is convenient because it saves energy and provides longer battery life. On the other hand, longer periods between queries increase the probability of an anomaly being undetected by a faulted cluster.

Finally, upon reception, the CC may take a few ms to raise the alarm. It must be noted, however, that once the alarm is raised, countermeasures must be taken within an interval such that they are effective, a time-constraint typical of real-time systems.

4 An EDEES to Detect Forest Fires

The general model just described will be applied now to the design of an EDEES slated to detect forest fires. The early detection of forest fires is a particularly important application since the forests burnt every year in the planet are in the order of hundreds of thousands of km^2, with heavy human life losses, extinction of species and aggravation of the greenhouse effect [11].

The system tries to detect natural conditions leading to the initiation of a forest fire and, therefore, anticipating the actual fire. In order to do that, a complex index with many sensed variables must be used. The Canadian Fire Weather Index (FWI) [12], is based on decades of research on the problem and takes into account ignition probability, fire behavioral characteristics in case it develops, difficulties to control it and the damages it could cause. FWI indicates fire intensity by combining the rate of fire spread with the amount of fuel being consumed.

The area to be monitored may be divided in grid squares. Since there are many sensed variables and the transmission range of the motes is only 30 m, a careful deployment of the nodes, in a cluster tree topology, has to be made in each square. In Figure 2, an example of a layout is given. The sides of the square are 120 m long, so its area is 0.0144 km^2.

It must be noted, however that the deployment presented in Fig. 2 is only theoretical because sensors are placed at the very limits of transmission ranges. It is dubious that such an extreme topology will work in practice. With this layout, the cluster-tree should be constructed as shown in Figure 3.

The infeasibility of covering the totality of large areas with WSNs has been pointed out in [8]. National Parks, for example, may have up to several thousand of km^2. In Argentina, for instance, the Nahuel Huapi National park has more than 7100 km^2. Even subtracting the area of the lake, about 500 km^2, the area to monitor fires remains large.

In order to have a grasp of the numbers involved, let's take as unit of study an area of 1000 km^2. To monitor 100% of it, 69.444 grid squares, as previously described, would be necessary, with a number of nodes near 2.5 million. The cost of equipment alone would be in the order of hundreds of millions of US dollars. To this, the cost of deployment must be added. Air sowing is impractical for several reasons: sensors may be caught by trees, they may not land right side up (bringing difficulties in transmission) and spacing may be incorrect. Setting them by hand is very labor intensive and may end adding an extremely heavy component to the final cost. What is worse, this is not a one-time-only expense since batteries die and the whole system must be replaced. Even if it is done only once every two years, the cost is extremely high. On top of that, the need of periodic replacements brings out the problem of millions of dead batteries polluting the soil.

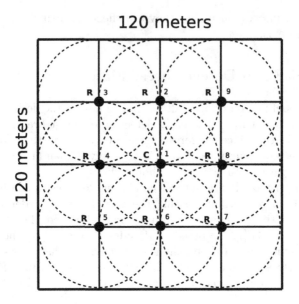

Fig. 2. Wireless sensor network deployment

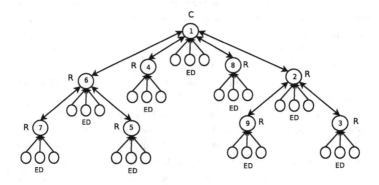

Fig. 3. Tree configuration

The deployment of the system is also unfeasible in small areas like, for instance, villages with houses densely built among groves. In this case, children playing in the lot, pets or wild animals may interfere with the nodes and render the system inactive.

Therefore, for the time being, WSNs do not seem to be ripe to implement EDEES covering large areas. They may be feasible in the future but, in the first place, they would require inexpensive motes with a longer range. The ratio *cost/range* should decrease by, at least, two orders of magnitude. In the second place, motes with a long life-span of their power supply will be necessary. This can be expected if, for instance, a replenishable supply of energy is obtained by scavenging the environment [13].

A second-best alternative is however possible, even with present motes' costs and technologies. Let's assume grid squares of 100 km^2, with sensors in the center. For each 1000 km^2 of park, there would be only 10 grid squares. The maximum distance between any point in the grid and the nearest C-node could be seen as the resolving power of the grid. It happens when the point is in the vertex of a grid square, and it would be approximately 7 km, a rather short distance considering that the grid square has an area of 100 km2.

Obviously, conditions leading to the starting of a fire would not be detected, but once the fire begins, it will progress and eventually reach the center of a square grid. Necessary sensors would be one for smoke, one for air temperature, one for wind direction and one for wind intensity, connected in a star topology. If the smoke sensor does not anticipate the fire, the air temperature sensor will detect an anomalous increase before being engulfed by the fire. Wind sensors will provide data allowing a good estimation of where and when the fire started. Adding sensors to calculate a complex fire index is useless since the covered area would be 0.0144% of the total area. Accepting four ED-nodes plus one C-node per grid square, a total of 50 nodes, with a cost of US$ 5000 at present prices, will satisfy the requirements of the 1000 km^2 of the study unit.

Obviously, the system does not give an early warning of an impending fire but an early notice that the fire is in its beginning phase. This could allow to take rapid countermeasures. This is in line with the opinion of professional fire-fighters reported in [14] indicating that a sensor that could indicate only the presence of fire would be useful even if the instrument was rapidly destroyed.

It must be noted that taking appropriate countermeasures, for instance sending an hydrant airplane followed by a fire-fighters team transported by cross-country 4WD terrestrial vehicles, may take, in both alternatives, a time several orders of magnitude bigger than the delay from sensing to alarm. Therefore, it is there where the efforts to reduce delays must be focused.

5   Conclusions

Presently, wireless sensor networks have been profusely proposed to monitor weather and field variables in places of difficult access. In this paper, a general system model based on a WSN sensing weather and field variables was proposed. Data is transmitted to a central computer, in charge of taking decisions, by IEEE 802.15.4 and GPRS as high and low end communication protocols. When applied to detect conditions leading to the initiation of forest fires it was shown that although from a theoretical point of view, the system may work well and meet the proposed targets, its deployment in large areas is deemed to be presently unfeasible because of the costs involved. A second best alternative, giving not an early warning of an impending fire but an early notice that the fire is already in its beginning phase, is shown to be not only feasible but also useful, from both a technical and an economic point of view.

References

1. Hefeeda, M., Bagheri, M.: Wireless sensor networks for early detection of forest fires. In: Proceedings of the 4th IEEE International Conference on Mobile Adhoc and Sensor Systems, Pisa, Italy (2007)
2. Son, B., Her, Y., Kim, J.G.: A design and implementation of forest-fires surveillance system based on wireless sensor networks for south korea mountains. IJCSNS International Journal of Computer Science and Network Security 6(9), 124–130 (2006)
3. Yu, L., Wang, N., Meng, X.: Real-time forest fire detection with wireless sensor networks. In: Proceedings of the 2005 International Conference on Wireless Communications, Networking and Mobile Computing, September 23-26, vol. 2, pp. 1214–1217 (2005)
4. Koubaa, A., Alves, M., Tovar, E.: Modeling and worst-case dimensioning of cluster-tree wireless sensor networks. In: RTSS 2006: Proceedings of the 27th IEEE International Real-Time Systems Symposium, Washington, DC, USA, pp. 412–421. IEEE Computer Society, Los Alamitos (2006)
5. Koubâa, A., Cunha, A., Alves, M., Tovar, E.: Tdbs: a time division beacon scheduling mechanism for zigbee cluster-tree wireless sensor networks. Real-Time Syst. 40(3), 321–354 (2008)
6. Koubaa, A., Alves, M., Tovar, E.: Gts allocation analysis in ieee 802.15.4 for real-time wireless sensor networks. In: IPDPS (2006)
7. Lu, G., Krishnamachari, B., Raghavendra, C.: Performance evaluation of the ieee 802.15.4 mac for low rate low power wireless networks. In: Workshop for energy efficient wireless communiucations and networks, EWCN 2004 (2004)
8. Tanenbaum, A., Gamage, C., Crispo, B.: Taking sensor networks from the lab to the jungle. Computer 39(8), 98–100 (2006)
9. Crossbow, http://www.xbow.com
10. Chen, X., Goodman, D.: Theoretical analysis of gprs throughput and delay. In: IEEE International Conference on Communications (2004)
11. Kuhrt, E., Knollenberg, J., Mertens, V.: An automatic early warning system for forest fires. Annals of Burns and Fire Disasters XIV(3) (2001)
12. Canadian, G.: Canadian forest fire danger rating system, http://www.nofc.forestry.ca/fire/
13. Rabaey, J.M., Ammer, J., Karalar, T., Li, S., Otis, B., Sheets, M., Tuan, T.: 12.3 pico radios for wireless sensors networks. the next challenge in ultralow power design. In: 2002 IEEE International Conference on Solid-State Circuits, ISSCC. Digest of Technical Papers, vol. 1, pp. 200–201 (2002)
14. Doolin, D., Sitar, N.: Wireless sensors for wildfire monitoring. In: Proceedings of SPIE Symposium on Smart Structures & Materials / NDE 2005, San Diego, USA (2005)

Deploying a Wireless Sensor Network in Iceland

Kirk Martinez[1], Jane K. Hart[2], and Royan Ong[3]

[1] School of Electronics and Computer Science, University of Southampton, SO17 1BJ, UK
[2] School of Geography, University of Southampton, SO17 1BJ, UK
[3] School of Engineering, Monash University, Malaysia
km@ecs.soton.ac.uk, jhart@soton.ac.uk, royan.ong@googlemail.com

Abstract. A wireless sensor network deployment on a glacier in Iceland is described. The system uses power management as well as power harvesting to provide long-term environment sensing. Advances in base station and sensor node design as well as initial results are described.

1 The Glacsweb Project

The Glacsweb project [1] aimed to study glacier dynamics through the use of wireless sensor networks. It replaced wired instruments which had previously been used with radio-linked subglacial *probes* which contained many sensors. The base of a glacier has a significant effect on a glacier's response to climate change and there is a growing need to study it in order to build better models of their behaviour. Several generations of systems were deployed in Briksdalsbreen, an outlet of the Jostedal icecap in Norway. As a multi-disciplinary research project it involved people from many domains: electronics, computer science, glaciology, electrical engineering, mechanical engineering and GIS.

Initial deployments had to solve the mechanical design of the probe cases and the unknown radio communication issues. The solutions involved craft as much as science and engineering but the key success has been to create data which had not existed before [2,3,4,5] while advancing our knowledge of sensor network deployments. Hot water drills are used in order to produce holes which reach the glacier bed. Most probes are placed 10-30cm under the ice while some are placed within the ice. Due to the relatively slowly changing environment the probe sense rate is normally set to once every four hours, although an adaptive sampling algorithm has been developed in the lab [6] which would optimise this sampling rate.

2 Iceland Deployment

Skalafellsjökull is a part of the large Vatnajökull icecap in Iceland and our site was chosen at (64°15'27.09"N, 15°50'37.68"W) around 800m altitude near an access road. Although there was no local internet connection there was a mobile phone signal which we used for the main internet link. The glacier is deep enough to test beyond 100m depth in the future (we used 60-80m).

Due to the nature of the team and time available a few key topics were chosen for the developments for the Iceland deployment. One main area was to improve the basestation

N. Trigoni, A. Markham, and S. Nawaz (Eds.): GSN 2009, LNCS 5659, pp. 131–137, 2009.

design through the use of a Gumstix processor. This would provide a better development environment and easier package management. The probes would maintain the PIC18 microcontroller but would gain an improved power supply and simplified code. In terms of sensors, a high resolution temperature sensor was required to sense small changes. A light reflectance sensor was also on the list of items to be tested in order to provide more information on the nature of the material surrounding the probes. A simple star network was used rather than our more complex TDMA-based protocol [7] in order to simplify debugging.

The disintegration of the ice front in Briksdalsbreen [8] meant that we lost our previous basestation infrastructure. This coupled with the higher altitude of the Iceland site (approximately 800m) led us to build a strong physical structure based on the previous pyramid design, to support antennas and the wind generator.

3 Probes

The sensor probe was developed around a low voltage version of a PIC18 microcontroller employing rigorous power management techniques, as shown in [9]. Power is supplied by three Lithium Thionyl Chloride cells chosen for their high energy density and wired in parallel. These cells constantly supply power to the real-time clock (RTC) and the 5V DC-DC converter. Three voltage regulators are used to significantly reduce power-rail noise due to the DC-DC converter, and increase the power supply rejection ratio (PSRR) between the digital and analogue circuits of the system.

The microcontroller is only powered when the RTC's alarm is triggered. The PIC controls the supply to the sensors and the wireless transceiver to avoid wastage of energy when they are not needed. A low drift (1ppm/°C) RTC was used to minimize the necessary synchronization window when the probes awaken each day for communication.

The sensor probes were configured to gather readings hourly. Apart from one reading, it powers off immediately after the sensors are read and the RTC's alarm is configured for the following wake up time. The sensor probe enables its transceiver after one daily reading at noon and remains powered for up to one minute to allow communication. A simple potential divider and opamp configuration used for the battery sensor allowed it to monitor the energy available to allow automatic schedule tuning. This rigorous power management scheme has been successfully employed in all versions of our sub glacial monitoring system, with the current design having a sleep current of 6μA (at 3.6V), and a daily consumption of approximately 550μWH.

The sensor probe uses the conductivity sensor to determine the presence of water within its immediate surrounding, and together with the reflectivity sensor to determine the presence of ice; we are able to infer the surrounding medium of the sensor probe: whether embedded it is in dry or wet till, or stuck in the bore hole. The strain gauge and pressure sensor measure the structural and hydrostatic (or atmospheric) pressure applied by the glacier, which themselves had been used in the past to infer the probe's surrounding medium.

The tilt sensors are integrated in a single micro-electro-mechanical system (MEMS) integrated circuit capable of measuring static acceleration in all orthogonal directions. This data is converted to the tilt/roll of the probe. The case strain is measured using strain gauges glued inside the case.

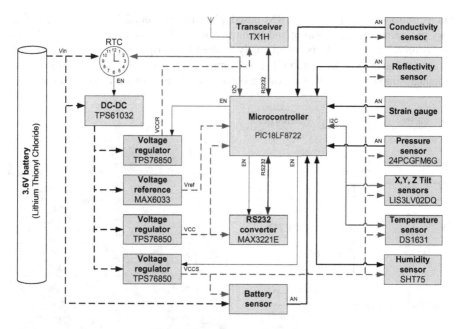

Fig. 1. Block diagram of a probe

The radio transceiver used was the BiM1-173.250-10 from Radiometrix which was tested in Norway. 173MHz was used rather than the more usual 433MHz, 868Mhz or 2.4GHz ISM bands due to the significant attenuation and reflections at higher frequencies. The antenna was a custom 1/4 wave helical design made from a 24 s.w.g. wire specifically tuned for transmissions through ice.

Each sensor probe has the provision for an RS232 level converter allowing it to be used as a "wired probe". In this mode of operation, the sensor probe acts as a distant transceiver which was embedded 10m within the glacier for improved communication range, particularly during the summer. The light sensor was a simple LED and photodiode configured with till/ice in the lab.

Calibration data for each probe is held on the basestation and the raw data is stored alongside a converted set of readings. This saves some effort by the probes and requires very little processing by the more powerful basestation. Twelve probes were deployed in the summer of 2008.

4 Base Station

Gateways or base stations are often a single point of failure in sensor networks. We have continually improved and redesigned a basestation every year since the first Norway deployment in 2001. The earliest were highly specialized and could not easily be reprogrammed remotely. The reality of deployments means that enhancements are often required after leaving the site. It is also very useful to be able to run experimental configurations for short periods of time. The 2008 basestation used research from the lab into combining a powerful processor running Linux with a low power microcontroller [6]. A small ARM-based platform produced by Gumstix provides easy programming, networking and management. In Norway we had decided that the main control code should be in an easily read/edited script language, with core code frozen (and generally written in C). The Gumstix platform made the use of the Python language simple and this was a major advance over Csh scripts previously used.

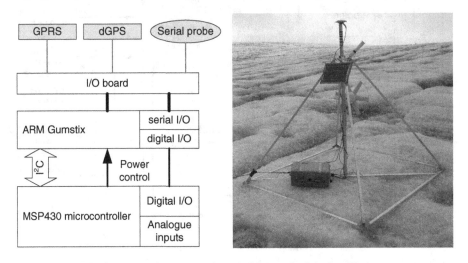

Fig. 2. Base station schematic and photograph of the installation

However one of the weaknesses shown by our previous ARM-linux component (Bitsy) was its high sleep current. The Gumstix (Connex) was not much better so the decision was made to power-off the ARM completely between uses and use the microcontroller (MSP430) to power it up. This "cold-boot" takes much longer than a standby wake-up but uses almost zero power. The energy wasted by booting for twenty seconds is compensated for by over 23hours in off-mode. This is made possible by the low frequency of basestation wake-ups we typically use – one per day. Clearly if such a system is woken too frequently the power saving benefits are lost. The microcontroller has a negligible sleep current (given that we operate from lead acid batteries) and can wake to carry out basestation sensing such as battery voltage, temperature etc.

As can be seen in Fig. 2, the base station structure is not anchored to the ice but has cutting edges and the weight of the base station "pelican" case is used to stabilise it. The Topcon dGPS antenna can be seen on the very top next to the slanted GPRS antenna. The wind generator is placed in a fixed position low down in order to increase stability. This was found to be a good compromise in Norway – as the wind tends to come down glacier and having a mobile generator would create stability problems. Figure 3 shows the case design – with its aluminium strengthening structure, which also separates the batteries in case of severe impacts. An internal case is always used in order to allow work when the weather is poor. It also allows the complete control unit to be removed easily. Large MIL-spec connectors are used on the outside of the Pelican case. A small round piezo sounder can be seen inside the internal case which

Fig. 3. Base station case showing major components

is used for debugging beeps. A USB network connection is used to connect a laptop in order to log into the Gumstix and upload/edit files or software. A compact flash card is used to create an archive of all data and logs. The compact flash card was one of the few components to survive falling into the lake in Norway (along with the Topcon GPS).

5   Initial Results and Analysis

In terms of system monitoring the solar and wind power worked well as shown by consistently high battery voltage readings. While it was known that the solar power would become negligible in winter the snow effects on the wind generator were unpredictable. The low power needed for the basestation meant we could compensate for this unknown by installing plenty of lead-acid battery capacity to maintain it through the winter (36AH). Sadly because of the GPRS loss in November we will have to wait to obtain the winter data. A problem with the wired probe's radio also

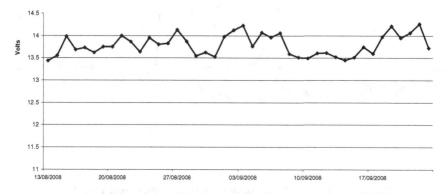

Fig. 4. Battery voltage of the base station showing consistently high charge after deployment

meant that after leaving the site no probe communications were possible. In Norway we had switched to four wired probes to remove this single point of failure. A "rescue mission" in 2009 will replace it as well as install other upgrades.

The base station computer was found to use 7µW in sleep mode, 52mW when taking sensor readings with the Gumstix off and 900mW when fully on (without GPRS or dGPS). It is the extremely low power in sleep mode which makes this the most frugal system we have produced.

The GPRS communications worked well even though the signal strength was not good and some of the daily files reached 110 kbytes. It was not possible to "tunnel" back into the system from the UK so the backup control of writing a "special" script which was fetched and executed daily worked well. This appears to be a weakness of using GPRS as an international network solution.

The data from the wired probe 12 shown in Fig. 5 shows that the light reflectance sensor is showing a change as the borehole closes around the probe. The temperature reading is a typical value for ice and the case strain (primary axis) is varying due to the ice closing around the case.

The dGPS was configured to record for five minutes once per week and these files would be correlated with those available from a national reference system in Höfn in order to measure ice velocity. While this leads to less accuracy due to the long baseline it saved us a task installing our own second dGPS. Long recordings were found to have an accuracy of 4mm because of the large numbers of satellites available. External data such as weather and a webcam image of the area are downloaded automatically by the server in the UK.

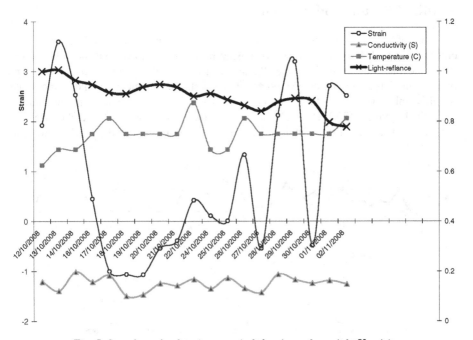

Fig. 5. Sample probe data (case strain left axis – others right Y axis)

6 Discussion and Conclusions

Every deployment has its own characteristics and issues which are only discovered once in the field. We built, programmed and deployed this system within four months, which is a record for us and was only possible through carefully cutting down the complexity of the system and because the team was technically extremely capable. The single point of failure in the wired probe turned out to be a weak point but we expect the probes to be still running, so gathering their stored data and re-establishing the network is still possible. The new base station architecture has proven a success, not only is it easier to develop with and uses less power but is also cheaper than our previous platform.

Future improvements will include adaptive GPS recording, so that in times of high power availability longer (hence more accurate) readings can be taken. A backup internet link is also needed, using long range modems to the nearest building.

Acknowledgements

The authors would like to thank the Glacsweb team for their dedicated efforts: Kathryn Rose, Jeff Gough, Robert Spanton, Tom Bennellick, Stuart Rimmer and James Cheshire. This research was funded by the EPSRC.

References

1. Martinez, K., Hart, J., Ong, R.: Environmental Sensor Networks. IEEE Computer 37(8), 50–56 (2004)
2. Hart, J.K., Rose, K.C., Martinez, K., Ong, R.: Subglacial clast behaviour and its implication for till fabric development: new results derived from wireless subglacial probe experiments. Quaternary Science Reviews (2009)
3. Rose, K.C., Hart, J.K.: Subglacial comminution in the deforming bed: inferences from SEM analysis. Sedimentary Geology 203, 87–97 (2008)
4. Hart, J.K.: An investigation of subglacial processes at the microscale from Briksdalsbreen, Norway. Sedimentology 53, 125–146 (2006)
5. Hart, J.K., Martinez, K., Ong, R., Riddoch, A., Rose, K.C., Padhy, P.: An autonomous multi-sensor subglacial probe: Design and preliminary results from Briksdalsbreen, Norway. Journal of Glaciology 51(178), 389–397 (2006)
6. Padhy, P., Dash, R.K., Martinez, K., Jennings, N.R.: A utility-based sensing and communication model for a glacial sensor network. In: 5th Int. Conf. on Autonomous Agents and Multi-Agent Systems, Hakodate, Japan (2006)
7. Elsaify, A., Padhy, P., Martinez, K., Zou, G.: GWMAC- A TDMA Based MAC Protocol for a Glacial Sensor Network. In: Proceedings of 4th ACM PE-WASUN 2007, Chania, Crete Island, Greece (2007)
8. Martinez, K., Basford, P., Ellul, J., Spanton, R.: Gumsense - a high power low power sensor node. In: 6th European Conference on Wireless Sensor Networks (2009)
9. BBC News at ten: 2007, http://www.youtube.com/watch?v=CY8AagMh_1M

Efficient Viewpoint Selection for Urban Texture Documentation

Houtan Shirani-Mehr, Farnoush Banaei-Kashani, and Cyrus Shahabi*

University of Southern California, Los Angeles, CA 90089, USA
{hshirani,banaeika,shahabi}@usc.edu

Abstract. We envision participatory texture documentation (PTD) as a process in which a group of participants (dedicated individuals and/or general public) with camera-equipped mobile phones participate in collaborative/social collection of the urban texture information. PTD enables inexpensive, scalable and high resolution urban texture documentation. PTD is implemented in two steps. In the first step, minimum number of points in the urban environment are selected from which collection of maximum urban texture is possible. This step is called *viewpoint selection*. In the next step, the selected viewpoints are assigned to users (based on their preferences and constraints) for texture collection. This step is termed *viewpoint assignment*. In this paper, we focus on the viewpoint selection problem. We prove that this problem is NP-hard, and accordingly, propose a scalable (and efficient) heuristic with approximation guarantee for viewpoint selection. We study, profile and verify our proposed solution by extensive experiments.

1 Introduction

The advent of earth visualization tools (e.g., Google Earth™, Microsoft Virtual Earth™) has inspired and enabled numerous applications. Some of these tools already include *texture* in their representation of the urban environment. The urban texture consists of the set of images/photos collected from the real environment, to be mapped on the façade of the 3D model of the environment (e.g., building and vegetation models) for photo-realistic 3D representation. Currently, urban texture is collected via aerial and/or ground photography (e.g., Google Street View). As a result, texture collection/documentation is 1) expensive, 2) unscalable (in terms of the required resources), and 3) with low temporal and/or spatial resolution (i.e., texture cannot be collected frequently and widely enough).

To address these limitations, we propose leveraging the popularity of camera-equipped mobile devices (such as cell phones and PDAs) for inexpensive and scalable urban texture documentation with high spatiotemporal resolution.

---

* This research has been funded in part by NSF grants IIS-0238560 (PECASE), IIS-0534761, CNS-0831505 (CyberTrust), and the NSF Center for Embedded Networked Sensing (CCR-0120778). Any opinions, findings, and conclusions or recommendations expressed in this material are those of the author(s) and do not necessarily reflect the views of the National Science Foundation.

N. Trigoni, A. Markham, and S. Nawaz (Eds.): GSN 2009, LNCS 5659, pp. 138–148, 2009.

With such *participatory texture documentation*, termed PTD hereafter, a group of participants (dedicated individuals and/or general public) with camera-equipped mobile phones participate in collaborative/social collection of the urban texture information[1]. By enabling low-cost, scalable, accurate, and real-time texture documentation, PTD facilitates various applications such as eyewitness news broadcast, urban behavior analysis, and real-estate monitoring as well as critical emergency-response and disaster management applications (e.g., in case of earthquake, hurricane, and wildfire).

PTD can be implemented as a two-step process. In the first step, called *viewpoint selection*, a set of points in the urban environment is selected for texture collection. We call such points *viewpoints*. Collection of the maximum possible urban texture should be doable by taking images at the set of selected viewpoints collectively. Because of the participatory nature of PTD, available resources (users' participation time) are usually limited and, therefore it is critical to minimize the number of selected viewpoints. In the second step, termed *viewpoint assignment*, considering the constraints and preferences of the participants, the selected viewpoints are assigned to different individuals for texture collection.

In this paper, we focus on the first step, viewpoint selection. We formally define the viewpoint selection problem and prove that it is an NP-hard problem by *reduction from* the minimum set cover problem [1]. Therefore, optimal solutions for the problem are rendered unscalable as the extent of the urban environment grows large; hence, we propose an efficient heuristic, termed GVS, with approximation guarantee to select the viewpoints. GVS solves a given instance of viewpoint selection problem by *reduction to* an instance of the minimum set cover problem. Based on our experimental results, as compared to the naïve approach which selects the environment points by imposing a grid with the cell size of $c \times c$, GVS reduces the number of selected viewpoints by 83% on average over different values of c.

The rest of this paper is organized as follows. In Section 2, we formally define the viewpoint selection problem. We study the complexity of the problem in Section 3. Section 4 discusses our approach to solve the viewpoint selection problem. Section 5 presents the results of our extensive empirical analysis of the proposed solution. Finally, we discuss the related work in Section 6, and conclude in Section 7.

2   Problem Definition

An urban environment consists of various 3D elements such as buildings, trees and the terrain (Fig. 1(a)).The environment can be modeled by any object-level model, i.e., a model in which the environment is represented by a set of objects. Here, without loss of generality, we assume a TIN (Triangulated Irregular Network) [2] model is used to represent all 3D elements in an urban environment (Fig. 1(b)). The texture of the environment is the set of images mapped on the triangles of the TIN model. Correspondingly, participatory texture documentation (PTD) is defined as the process of collecting and mapping the texture onto the TIN model of the urban environment.

---

[1] A PTD system is currently under development at the Information Laboratory in the University of Southern California (see http://infolab.usc.edu/projects/GeoSIM).

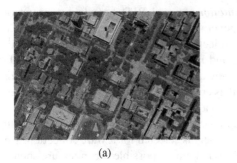
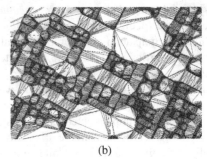

(a)   (b)

Fig. 1. Representing an urban environment for texture documentation: (a) The environment and (b) its representation in TIN model

With PTD, participants/users move in the urban environment and make stops at selected viewpoints to take images. We assume users' movements are limited to an urban road network and, therefore, the viewpoints must be selected from the road network. The road network can be modeled by a weighted graph G, which we call the *road network graph* hereafter. We assume a user walks or drives on G and she can stop and take images at only certain points on G, denoted by P. Users are presumed to carry similar camera-equipped cell phones. A user takes a panoramic image at an assigned viewpoint v on the road network and therefore collects the maximum texture one can collect at v.

Given the aforementioned assumptions, we define the viewpoint selection problem as follows. Consider an environment E represented by TIN model with T as the set of TIN triangles, and a road network graph G in E. Assume the set of points on G at which a user can make stop and take images is denoted by P. We define $T' \subseteq T$ as the TIN triangles whose texture can be collected from the points in P (note that viewing the texture of all the triangles in T might not be possible from P, because P is only a limited subset of all possible viewpoints in E). Accordingly, we call a set $S \subseteq P$ a *texture covering* set, if collection of texture for *all* the triangles in T' is possible by taking images at the points in S. The viewpoint selection problem is defined as the process of finding a texture covering set V with minimum size. Fig. 2 illustrates this process. Fig. 2(a) shows the original road network graph G. In Fig. 2(b), the selected viewpoints on G are shown as circles. For two viewpoints vp_1 and vp_2, the areas whose texture can be collected at either of them are highlighted.

3   Complexity Result

In this section we prove that the viewpoint selection problem is NP-hard by reduction from the *minimum set cover* problem [1]. We first state the minimum set cover problem.

Definition 1. *Let* $S = \{s_1, s_2, \ldots, s_m\}$ *be a collection of finite sets,* s_i's *, whose elements are drawn from a universal set* U. *Let* $F = \bigcup_{i=1}^{m} s_i$ *where* $F \subseteq U$. *Minimum set cover finds a set* C *with minimum cardinality where* $C \subseteq S$ *and* $\bigcup_{s \in C} s = F$.

(a)   (b)

Fig. 2. Viewpoint selection: (a) Original road network graph (vertices are shown as squares); (b) Road network after viewpoint selection (viewpoints are shown as circles)

For example, assume $U = \{1, 2, 3, 4, 5, 6\}$ and $S = \{s_1, s_2, s_3\}$, where $s_1=\{1, 2\}$, $s_2=\{2\}$ and $s_3=\{1, 3\}$. Thus, $F = \{1, 2, 3\}$ and the minimum set cover is $C = \{s_1, s_3\}$. The minimum set cover problem is NP-hard. The following theorem proves that the viewpoint selection problem is also NP-hard.

Theorem 1. *The viewpoint selection problem is NP-hard.*

Proof. We prove the theorem by providing a polynomial time reduction from the minimum set cover problem. Towards that end, we prove that given an instance of the minimum set cover problem, denoted by SCI, there exists an instance of the viewpoint selection problem, denoted by VSI, such that the solution to SCI can be converted to the solution of VSI in polynomial time. Consider a given SCI having U as the universal set, $S = \{s_1, s_2, \ldots, s_m\}$ where $s_i \subseteq U$, and let $F = \bigcup_{i=1}^{m} s_i$. To solve SCI, we select a set $C \subseteq S$, with minimum cardinality, to cover all the elements in F. Correspondingly, to solve a VSI, we look for a $V \subseteq P$, with minimum cardinality, such that collection of texture for all the triangles in T' is possible from the points in V. Therefore, we propose the following mapping from SCI components to VSI components to reduce SCI to VSI. Suppose the universal set U corresponds to the set of triangles T. Each $s_i \in S$ is mapped to a point $p_i \in P$ as selection of sets in SCI corresponds to selecting the viewpoints in VSI. Finally, we map F to T'. The intuition behind the last mapping is that with SCI we want to cover each element $t \in F$ and accordingly we aim to cover the triangles of T' in VSI with texture. We next explain each mapping in detail.

For mapping S to P, we assume there exist a point p_i in E corresponding to $s_i \in S$. A road network may exist to connect the points in P. Next, we assume a triangle $tr_j \in T'$ exists corresponding to $t_j \in F$. tr_j is visible to $p_i \in P$ if and only if $t_j \in s_i$ and is invisible to the other points because of existence of obstacles in E. It is easy to observe that if the answer to VSI is the set V, the answer to SCI will be the set $C = \{s_i | p_i \in V\}$. This completes the proof. $\square$

Based on the above theorem, the viewpoint selection problem is NP-hard, which makes the optimal algorithms impractical. In the next section, we provide an efficient heuristic with approximation guarantee for the viewpoint selection problem.

4 Efficient Viewpoint Selection

Our proposed heuristic algorithm, termed Greedy Viewpoint Selection (GVS), is based on reduction to (not reduction from) the minimum set cover problem. Such a reduction enables us to adapt the existing algorithms for minimum set cover problem to solve viewpoint selection problem. Here, we first explain our proposed reduction, and thereafter describe our GVS algorithm.

4.1 Reduction

For each point $p \in P$, we define the *visibility* set $vs(p)$ as the set of triangles in T' that are visible from p; i.e., $vs(p) = \{tr \in T'|V(p, tr) = 1\}$, where $V(p, tr) = 1$ if tr is visible from p. Without loss of generality, we assume a triangle tr is visible from p if every point of tr is visible from p. For example with $vs(p) = \{tr_1, tr_2, tr_3\}$ the triangles tr_1, tr_2 and tr_3 are visible from p. Therefore, a user standing at p can collect texture for these triangles. We construct the corresponding instance of the minimum set cover problem as follows. We define $F = \bigcup_{p \in P} vs(p)$ and $S = \{s = vs(p)|p \in P\}$. The universal set U can be any set such that $F \subseteq U$. For example, for an instance of viewpoint selection problem in which $P = \{p_1, p_2\}$, $vs(p_1) = \{tr_1, tr_2\}$, and $vs(p_2) = \{tr_3\}$, the corresponding set cover instance has $F = \{tr_1, tr_2, tr_3\}$, $S = \{vs(p_1), vs(p_2)\}$ and $U = F$. If the answer to the constructed minimum set cover instance is C, then the answer to the original viewpoint selection can be derived as V, where $p_i \in V$ if and only if $vs(p_i) \in C$.

Consequently, we can use any of the heuristics for the minimum set cover to solve viewpoint selection problem. To develop our viewpoint selection algorithm (see Section 4.2), we are inspired by the greedy minimum set cover algorithm from [3], which has a linear running time of $O(\sum_{i=1}^{m} s_i)$ where $s_i \in S$. The greedy minimum set cover algorithm works by iteratively selecting the set $s_i \in S$ that covers the greatest number of remaining uncovered elements of F. This algorithm is guaranteed to result in a suboptimal answer with an approximation guarantee of $\ln(n)+1$, where n is the cardinality of the set $s_j \in S$ with the largest number of elements.

4.2 Greedy Viewpoint Selection (GVS)

Before presenting our viewpoint selection algorithm, we define our terminology. *Texture score* is a measure which represents the amount of texture that can be collected from viewpoints. A user standing at a point p can collect texture for all the triangles visible from p. Correspondingly, the texture score of p, denoted by $TS(p)$, is the number of triangles visible from p; i.e., $TS(p) = |vs(p)|$. Similarly, the texture score of a set of points S, $TS(S)$, is defined as the number of triangles visible to *any* point in S; i.e., $TS(S) = |vs(S')|, S' = \bigcup_{p \in S} vs(p)$.

The pseudocode of our viewpoint selection algorithm, termed Greedy Viewpoint Selection (GVS), is presented in Algorithm 1. The algorithm takes as input the triangles in $T' \subseteq T$ and the set of road network points P. We explain the logic of the algorithm as follows. First, GVS computes the visibility set of each point $p_i \in P$ (lines $3 - 9$). Thereafter, with a greedy approach the viewpoint p with maximum texture score is iteratively selected among all points in P, removed from P, and added to the result set

V (lines $11 - 15$). Note that once p is added to V the corresponding triangles visible to p are excluded from T' (because they are covered); consequently, the texture score of V (and P) is updated at each iteration. The iteration correctly terminates when V becomes a texture covering set, i.e., $TS(V) = collectableTexture$.

To prove the correctness of GVS, note that all the triangles in T' can be texture mapped by at least one point in P. During each iteration, a point is selected and all the triangles visible to it are excluded. Therefore, the number of remaining triangles which cannot be texture mapped from the already selected viewpoints decrease, and correspondingly $TS(V)$ increases until the iteration terminates. The returned set V is the approximate answer to the viewpoint selection problem as it may have larger cardinality than the optimal viewpoint selection answer. Assuming that the optimal answer is V_{opt}, the size of the answer returned by GVS satisfies the following inequality:

$$\frac{|V|}{|V_{opt}|} \leq \ln(TS_{max}) + 1,$$

where TS_{max} is the texture score of the point with the largest texture score in P. The above inequality guarantees that the size of V is at most $\ln(TS_{max})+1$ times larger than the size of the optimal answer. This bound follows from the approximation guarantee of the greedy minimum set cover algorithm.

Algorithm 1. $GVS(T', P)$

1: $V = \emptyset$;
2: $vs = \emptyset$;
3: **for** $i = 1$ to $|P|$ **do**
4:    **for** $j = 1$ to $|T'|$ **do**
5:        **if** $V(p_i, t_j) = 1$ **then**
6:            $vs(p_i) = vs(p_i) \cup t_j$;   {Visibility set of each point is set}
7:        **end if**
8:    **end for**
9: **end for**
10: $collectableTexture = |T'|$
11: **while** $TS(V) < collectableTexture$ **do**
12:    $p = MaxTS(P)$;   {$MaxTS(P)$ returns $p \in P$ with maximum texture score}
13:    $V = V \cup p$;   {p is added to the answer set}
14:    $T' = T' - vs(p)$;   {T' is updated}
15: **end while**
16: **return** V;

---

5   Experiments

In this section, we evaluate our proposed solution by extensive empirical analysis. We first describe our experimental setting and then present the experimental results.

5.1   Experimental Methodology

We conducted our experiments using two real-world datasets. The first dataset, LA dataset, is the elevation data of Los Angeles area, from USGS (http://data.geocomm.com), covering a $10km \times 10km$ area. The second dataset, USC dataset, is the elevation data of University of Southern California campus covering a $1.5km \times 1.5km$ area. The road network data for LA dataset is acquired from NAVTEQ (http://www.navteq.com). For USC dataset, we assume all points with elevation zero comprise the road network; hence, the road network includes both roads and sidewalks.

To generate the TIN model for the urban environments covered by the datasets, first a set of ground positions with 3D coordinates among the total of approximately $180,000$ $(120,000)$ points are sampled from LA (USC) dataset, and subsequently Delaunay Triangulation [4] is used to generate the TIN model. We can change the number of triangles and consequently the resolution of the texture to be mapped by changing the number of sampled ground points. Increasing (decreasing) the number of samples will increase (decrease) the number of triangles and hence the resolution of the texture to be mapped. The number of samples and that of the corresponding TIN triangles for the two datasets are shown in Fig. 3. We selected fewer triangles to represent the USC dataset in TIN model as it covers a smaller area as compared to the LA dataset.

Moreover, to quantize the road network space, we impose a grid on the road network and use the intersection of road network segments and grid cells as the collection of road network points. This approach enables us to emulate various viewpoint selection restrictions by changing the granularity of the imposed grid. We assume users can stand at any of the resulted road network points. In our experiments we imposed grids with different granularity. The number of resulting points P by imposing different grids is shown in Fig. 4 for each dataset. We denote the set P generated by imposing a grid with the cell size of $c \times c$ meters as P_c. Finally, with all of our experiments we assume a point further than 400 meters to a point p is invisible from p.

Our experimental system is implemented in Java, and runs on a typical Intel 2.66GHz PC with 3.25GB RAM. The operating system is Windows XP SP2. For each setting,

Number of Samples	2000	4000	6000	8000	10,000	20,000	50,000
Number of Triangles	4000	8000	12,000	16,000	20,000	40,000	100,000

(a) LA

Number of Samples	2000	3000	4000	5000	6000	7000
Number of Triangles	4000	6000	8000	10,000	12,000	14,000

(b) USC

Fig. 3. Number of samples vs. number of triangles for different datasets

Grid	P_{20}	P_{40}	P_{60}	P_{80}	P_{100}
$|P|$	46303	11569	5152	2963	1825

(a) LA

Grid	$P_{2.5}$	P_5	$P_{7.5}$	P_{10}	$P_{12.5}$
$|P|$	57056	14317	6362	3637	2295

(b) USC

Fig. 4. Different values of $|P|$ in our experiments

we tested the algorithm by running it 10 times to compute the average values. Next, we present our experimental results.

5.2 Experimental Results

Based on our experiments, the optimal algorithm for viewpoint selection problem takes more than a day for $|P|$=25 (i.e., when P includes 25 randomly picked points from the road network imposed on LA dataset) and also its running time increases exponentially which makes it impractical. We observed that for $|P| < 25$, $\frac{|V|}{|V_{opt}|} \geq 95\%$, where $|V|$ ($|V_{opt}|$)is the number of viewpoints calculated by GVS (optimal) algorithm. Therefore, GVS clearly outperforms the optimal algorithm in efficiency and scalability, while providing almost optimal results. In this section, we study the effect of different parameters on the performance of GVS algorithm.

5.2.1 Size of Selected Viewpoints
With this experiment, we evaluate the effect of resolution on the number of selected viewpoints, i.e., $|V|$. The result is shown in Fig. 5 where each curve is generated for a specific P_c. Using GVS, on average $|V|$ is 83% (90%) less than $|P|$ for LA (USC) dataset over all the cases. This reduction significantly improves the scalability of any viewpoint assignment algorithm. As expected, as the number of samples grows $|V|$ increases, because the number of triangles which must be texture mapped increases. Similarly, increasing the number of road network points results in larger $|V|$, because more triangles become visible to the road network viewpoints.

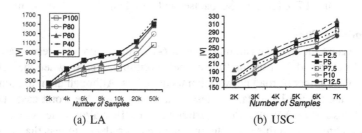

(a) LA   (b) USC

Fig. 5. $|V|$ vs. number of samples

5.2.2 Collected Texture
Viewing the texture for all the triangles in T (the collection of all the environment triangles) might not be possible from the points in P (see Section 2). Here, we measure the percentage of texture which can be collected from a set of selected viewpoints. This percentage is denoted by $\rho = \frac{TS(V)}{|T|}$. For example, $\rho = 50\%$ states that only 50% of the environment triangles can be texture mapped by taking images at the viewpoints. Fig. 6 illustrates how ρ varies for different number of samples and road network points. For both datasets, increasing the number of samples increases the value of ρ since smaller triangles will be introduced and the chance that a triangle is visible to a point increases. The increase rate of ρ for USC dataset is slower as a large portion of this dataset is the

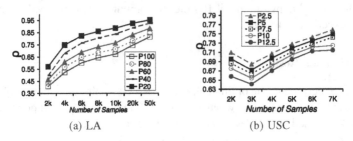

Fig. 6. ρ vs. number of samples

ground points with the same elevation of zero for which increase in the resolution by raising the number of samples does not have much effect.

For a fixed number of triangles, increasing the number of road network points raises the value of ρ as more texture can be collected from more road network points. The maximum value of ρ for the LA dataset is $\rho = 96\%$ which is obtained with 50000 samples and P_{20}. For the same number of samples if we choose P_{40}, ρ drops to 93%. This means that having much smaller P ($\approx 75\%$ decrease in the number of road network points) we can collect almost the same amount of texture. The reason is that for the finest grid , i.e., P_{20}, the distance between a point and its neighbor is very small and hence their visibility sets are almost the same. Therefore, the amount of texture which can be collected is not much more than the coarser grid of P_{40}. Based on our experiments and for a specific number of samples, the difference between the value of ρ for different number of road network points is at most 30% for the LA dataset. This difference is at most 7% for the USC dataset and less than 5% for 7000 samples.

5.2.3 Running Time

We also measure the average running time of GVS algorithm. Our results are shown in Fig. 7. As expected the running time of the algorithm increases when the number of samples or the number of road network points increases. Although the maximum value of running time for the LA dataset is less than 7 minutes in the worst case (i.e., with 50000 samples and P_{20}) as mentioned earlier we can collect almost the same amount of texture by having P_{40}. In this case, the running time is reduced to less than 3 minutes. Similarly, for the USC dataset the running time is reduced to less than 2 minutes when using $P_{7.5}$ instead of $P_{2.5}$ which results in collection of only 5% fewer texture.

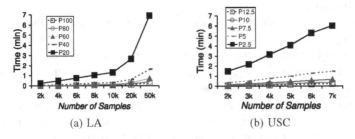

Fig. 7. Running time vs. number of samples

6   Related Work

Texture mapping based on the images acquired by cameras is extensively studied in the literature [5,6]. For example, in [5] a system is developed to generate texture of building exteriors from mosaics of close-range photographs acquired with commodity digital cameras. Although with our work we also use the images taken by users to generate texture information, our focus is on selecting the minimum number of points in the environment from which gathering the complete texture of the environment is possible.

A second body of relevant work is the literature on the sensor deployment and sensing coverage with sensor networks. In [7,8], the coverage problem is formulated as a decision problem to determine whether every point in the service area of the sensor network is covered by at least k sensors. On the other hand, with sensor deployment the goal is to maximize the coverage by proper sensor placement, somewhat similar to our viewpoint selection problem. However, most of the proposed approaches for sensor deployment assume simple sensing models with circular (omnidirectional or unidirectional) coverage for sensors [9,10,11]. While these models properly approximate the coverage of the sensors with typical sensing modalities (e.g., sound and temperature sensors), visibility coverage is more complex; hence, rendering these approaches inapplicable for visual sensor deployment. The most relevant work to our work is on visual sensor deployment [12,13]. The art gallery problem is a classic work in this category [12]. Our work extends this category by studying the coverage at the object level, where objects can be modeled by the TIN model (we emphasize that the choice of TIN model in our problem setting is arbitrary and other object models can be used equivalently to represent the urban environment). Moreover, we consider spatial restrictions in viewpoint selection (e.g., selected points must be on a road network). Most importantly, while the previous work has focused on solving the visual sensor deployment problem in a continuous space, with GVS we assume a discrete space for viewpoint selection to model restrictions on where users can take images.

7   Conclusion and Future Work

In this paper, we introduced and studied the problem of viewpoint selection for participatory texture documentation. We studied the complexity of the problem, and since the problem is NP-hard we proposed GVS as an efficient heuristic solution with approximation guarantee. Moreover, we showed the efficiency of GVS empirically by extensive experiments. As part of our future work, in short term we intend to extend GVS to consider the visual "quality" of the texture in addition to texture coverage. The visual quality of the texture can be affected by distance and/or view angle. This will also require a fuzzy setting for our texture-score measure. In long term, we plan to address the viewpoint assignment problem.

References

1. Chakravarty, S., Shekhawat, A.: Parallel and serial heuristics for the minimum set cover problem. J. Supercomput. 5(4), 331–345 (1992)
2. Fowler, R.J., Little, J.J.: Automatic extraction of irregular network digital terrain models. In: SIGGRAPH 1979, pp. 199–207. ACM, New York (1979)

3. Cormen, T.H., Leiserson, C.E., Rivest, R.L., Stein, C.: Introduction to Algorithms, 2nd edn. McGraw-Hill Science/Engineering/Math (2001)
4. Chen, G.H., Yu, M.S., Liu, L.T.: Two algorithms for constructing a binary tree from its traversals. Inf. Process. Lett. 28(6), 297–299 (1988)
5. Tsai, F., Lin, H.C.: Polygon-based texture mapping for cyber city 3D building models. Int. J. Geogr. Inf. Sci. 21(9), 965–981 (2007)
6. Guillou, E., Meneveaux, D., Maisel, E., Bouatouch, K.: Using vanishing points for camera calibration and coarse 3D reconstruction from a single image. The Visual Computer 16(7), 396–410 (2000)
7. Huang, C.F., Tseng, Y.C.: The coverage problem in a wireless sensor network. In: WSNA 2003 (2003)
8. Meguerdichian, S., Koushanfar, F., Potkonjak, M., Srivastava, M.B.: Coverage problems in wireless ad-hoc sensor networks. In: INFOCOM, pp. 1380–1387 (2001)
9. Wu, C.H., Lee, K.C., Chung, Y.C.: A delaunay triangulation based method for wireless sensor network deployment. Comput. Commun. 30(14-15), 2744–2752 (2007)
10. Dhillon, S.S., Chakrabarty, K.: Sensor placement for effective coverage and surveillance in distributed sensor networks 3, 1609–1614 (2003)
11. Guestrin, C., Krause, A., Singh, A.P.: Near-optimal sensor placements in gaussian processes. In: ICML 2005 (2005)
12. Lee, D.T., Lin, A.K.: Computational complexity of art gallery problems. IEEE Trans. Inf. Theor. 32(2), 276–282 (1986)
13. Hörster, E., Lienhart, R.: On the optimal placement of multiple visual sensors. In: VSSN 2006 (2006)

User Requirements and Future Expectations for Geosensor Networks – An Assessment

Lammert Kooistra[1], Sirpa Thessler[2], and Arnold K. Bregt[1]

[1] Centre for Geo-Information, Wageningen University, 6700 AA
Wageningen, The Netherlands
{Lammert.Kooistra,Arnold.Bregt}@wur.nl
[2] MTT Agrifood Research Finland, Luutnantintie 13, 00410 Helsinki, Finland
sirpa.thessler@mtt.fi

Abstract. Considerable progress has been made on the technical development of sensor networks. However increasing attention is now also required for the broad diversity of end-user requirements for the deployment of sensor networks. An expert survey on the user requirements and future expectations for sensor networks was carried out. Both technology and applications are seen as main drivers for sensor network deployment however harmonization of (open) standards to collect, access, manage, and integrate sensor data are considered crucial for further development. Although sensor based applications are increasingly used in every day life, their use in decision making requires further improvement of aspects like privacy, data quality, etc. Finally, next to formal sensor networks, standardization will allow voluntary sensor information to become a significant sensor data source.

Keywords: Sensor network, sensor web, spatial data infrastructure, interoperability, sensor integration.

1 Introduction

To facilitate monitoring of a changing world, integrated information systems are required which are capable of real-time measurement of fundamental processes in the environment, as well as providing vital hazard warnings. Traditionally, sensor networks covering various geographical and temporal scales are an important source for this task [1]. They allow vast amounts of relevant information to be collected with a high temporal frequency for a network of point locations that are remote, inaccessible, or lack the necessary resources to acquire such information in a different manner. To date, much of the current sensor network research is technology-driven, focusing on applying and integrating a broad range of both experimental and off-the-shelf technology. Although novel and useful deployments have been demonstrated in a wide range of applications and environmental conditions [2], scale-ability still needs to be tested in most cases. From a hardware perspective, problems with energy consumption and communication bandwidth and reliability remain important issues to be solved. From the perspective of computer science, challenges focus on creating intelligent, robust, and self-healing networks, and creating easy access to sensors and their

N. Trigoni, A. Markham, and S. Nawaz (Eds.): GSN 2009, LNCS 5659, pp. 149–157, 2009.

data. From an application perspective, new ways of collecting and representing the data can help us to better understand and manage our changing world.

Although much progress has been made on the technical part of sensor networks, increasing attention is now also required for the broad diversity of end-user requirements for the deployment of sensor networks. Examining the chain from the end-user to the individual sensor nodes requires a framework to assess different implementation aspects not only including technology. This identified data management challenge for geosensor network applications could benefit from an examination of the history of and current developments in Spatial Data Infrastructures (SDIs). SDIs facilitate and coordinate access, exchange and sharing of spatial data at different spatial scale levels, between a wide variety of users in an easy and secure way. Although SDIs have been constantly evolving through the adaptation to changing circumstances and interests of heterogeneous actors [3], five interrelated aspects are commonly recognized as the main drivers for the development of SDIs: standards, policies, technology, people and data [4]. Latest definitions of the second generation of SDIs also put more emphasis on the inclusion of services, serving as a starting point and gateway to the web for a user community [4]. In this concept open standards are used for the management and exchange of geospatial data and the realization of service-oriented architectures aiming at the achievement of true interoperability [5]. This closely relates to the concept of sensor webs as introduced by Delin in 2002 [6], which envisions the development of a framework of open standards for exploiting web-connected sensors and sensor systems of all types. Recent activities within the Open Geospatial Consortium (OGC) have resulted in a suite of standards called Sensor Web Enablement (SWE) which include access to sensor measurements, retrieval of sensor metadata, controlling sensors, alerting based on sensor measurements and automatic processing of sensor measurements [7].

In this paper we present the results of an expert survey on the user requirements and future expectations for geosensor networks from an end-user perspective. In analogy to SDIs, we used the identified aspects: standards, policy, technology, people and data, as a starting point for the assessment of the field of sensor networks. The main drivers and limitations for the individual aspects will be discussed and (research) requirements for future sensor network development will be elaborated.

2 Survey Methodology

To assess the development of sensor networks and to identify their driving factors an expert survey was conducted. The survey was part of the workshop entitled "Sensing a Changing World" organized at Wageningen University in the Netherlands in November 2008 (http://www.grs.wur.nl/UK/workshops/scw). The workshop brought together 56 participants from 16 countries to discuss progress, opportunities, and challenges in sensor network research from a user perspective. The respondents considered themselves as technology and application developers and came from engineering, physics, environmental, geo-information and computer sciences. They used a range of operational definitions of sensors, sensor networks and sensor webs.

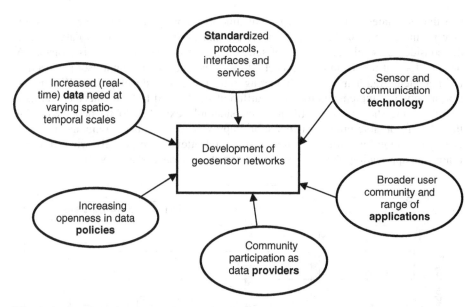

Fig. 1. Assumed driving factors for the development of geosensor networks. The bold terms describe the overlap with identified factors for the development of Spatial Data Infrastructures (SDIs) [4].

The expert survey consisted of four modules:

1. Identification of background, application field and experience of users with sensor network research;
2. Identification of main driving factors for the development of sensor networks at this moment. The respondents needed to rate statements in relation to the factors as indicated in Fig. 1. on a scale between 1 (totally disagree) and 7 (totally agree);
3. Identification of detailed issues per factor as presented in Fig. 1. For every factor, six statements needed to be rated on a scale between 1 (totally disagree) and 7 (totally agree);
4. View on future deployment of sensor networks with special emphasis on the development from sensor networks to sensor web based approaches. The respondents were asked to prioritize most important benefits and bottlenecks and give their view on this development.

A total of 22 participants answered all questions of the survey. The responses were not stratified according to specific characteristics like experience with sensor networks but were treated as one dataset. The complete survey consisted of 50 questions and statements; in this paper the most important results will be presented.

3 Results and Discussion

Rapid development of sensor and communication technology is rated as the main driving factor for evolvement of sensor networks (Fig. 2). This typically relates to a technology push process, however also user pull aspects like open standards, applications and the associated data needs (e.g., real time availability) are considered important drivers. Increasing number of data providers and users and opening of data policies are considered as relevant drivers of sensor network development at this moment. A detailed look at the technology aspects (Fig. 3) indicates that from a user perspective technological fragmentation is hindering sensor network development. Within the sensor network implementation chain different fields need to be connected and, as a result, off-the-shelf sensors still need considerable expertise to be deployed. Although considerable progress is made on improvement of sensor communication and energy-consumption, still this requires structural attention, especially when in environmental applications sensors are used in remote and inaccessible locations.

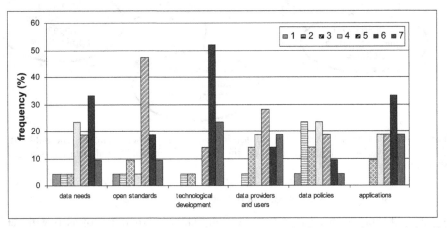

Fig. 2. Rating for assumed driving factors of geosensor networks. Rates vary between 1 (totally disagree) and 7 (totally agree).

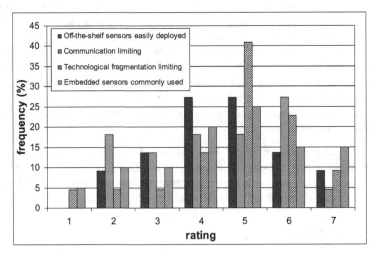

Fig. 3. Detailed statements for technology factor in relation to the development of geosensor networks. Rates vary between 1 (totally disagree) and 7 (totally agree).

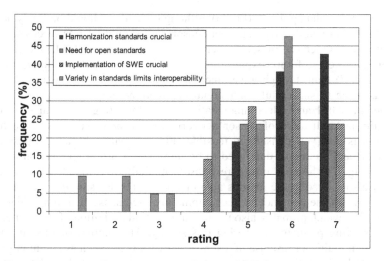

Fig. 4. Detailed statements for standardization factor in relation to development of geosensor networks. Rates vary between 1 (totally disagree) and 7 (totally agree).

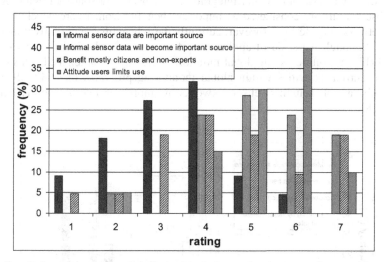

Fig. 5. Detailed statements for people factor in relation to development of geosensor networks. Rates vary between 1 (totally disagree) and 7 (totally agree).

Harmonization of standards was rated the highest of all statements and considered a crucial aspect for further development of sensor networks (Fig. 4). In addition, these standards should be based on open sources and the sensor web enablement standards as developed by the Open Geospatial Consortium (OGC) [7] were seen as a relevant option for further development. To achieve increasing availability of informal sensor data from voluntary networks, which is anticipated as future development (Fig. 5), standardization will be the key to access and retrieve data from a wide variety of sensors. However, the attitude of users (e.g., mistrust against data providers, data quality) could limit the utilization of both formal and informal sensor data sources.

With regard to data policies (Fig. 6), although metadata should be freely accessible, there is no overall agreement that sensor data should be free of costs. At this moment costs hinder use of sensor networks mainly as a result of high initial investments, however, future exploration is seen more through profitable business models (Fig. 7) than through governmental funding (Fig. 6). In this respect, reference was made to successful commercial applications like car navigation systems that combine GPS and telecommunication technology (e.g., for real time traffic information). Application of sensor networks in decision making without human interaction is not (yet) considered feasible (Fig. 7), mainly because robustness of the systems and quality of data need to be improved. Current application in everyday life (Fig. 7) shows a mixed response with half of respondents disagreeing, probably also related to the reasons also mentioned for decision making. However, the other half of the respondents sees commonly used sensor based applications in everyday life, which probably also relates to the broad definition of sensor networks taken as starting point for answering this survey. The support of several applications through multi-purpose use of a sensor network is not seen as common practice (Fig. 7). Probably in the current situation this is often restricted by data policy or unavailability of services for accessing the data.

Real-time data availability and integration of sensor data sources (including e.g., earth observation) are considered as important benefits from future development of sensor networks (Table 1). However, currently these benefits are not fully utilized yet (Fig. 8). Although the scale of application of sensor networks still remains local to regional (Fig. 8), future use in global monitoring systems is anticipated. This closely relates to current activities within global monitoring programs like GEOSS, which specifically aim at including sensor networks by improving interoperability between systems [8].

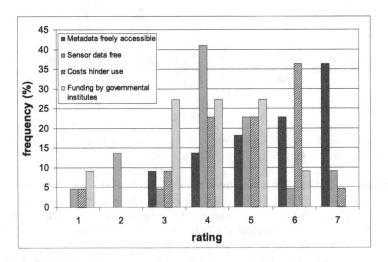

Fig. 6. Detailed statements for data policies factor in relation to development of geosensor networks. Rates vary between 1 (totally disagree) and 7 (totally agree).

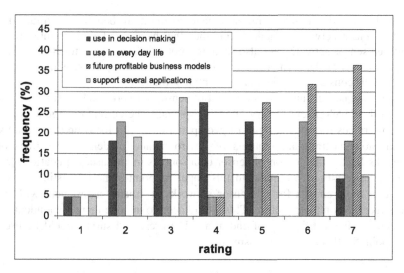

Fig. 7. Detailed statements for applications factor in relation to development of geosensor networks. Rates vary between 1 (totally disagree) and 7 (totally agree).

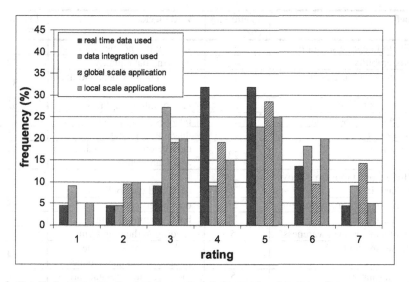

Fig. 8. Detailed statements for data factor in relation to development of geosensor networks. Rates vary between 1 (totally disagree) and 7 (totally agree).

The final part of the survey focused on the future development of sensor networks with special emphasis on the opportunities of sensor web based applications. In this context a sensor web was defined as "an autonomous instrument for collecting, storing, sharing and analyzing real-time sensor data which is open for all and is interoperable with other sensor webs/networks through open standard interfaces and protocols". Respondents considered the ability to integrate sensor data combined with interoperability and continuous measurements as the main benefits for future development of sensor networks (Table 1). On the other hand, limitations in interoperability between sensor networks are also seen as a potential bottleneck together with data quality and costs of deployment and maintenance. However, the sensitivity and privacy of data is considered as the main potential bottleneck.

As indicated earlier future development of sensor networks and sensor webs could reflect on the development history of SDI. Especially the current emergence of standard web services for diverse types of geo-spatial data with focus on (real time) sensor data (now casting) and outputs of environmental models (forecasting) offer benchmarks for the integration of sensor data sources in end-user applications [5, 9]. Although the results as presented in this paper pinpoint some interesting views on current user requirements for geosensor networks, a broader assessment will be required to get the complete picture. This would need to include more respondents with special emphasis on the group of end-users. The presented survey could be used as starting point for this broader assessment.

Table 1. Main benefits and bottlenecks for future deployment of sensor networks and development into sensor web based applications

Benefit	Frequency (%)	Bottlenecks	Frequency (%)
ability to integrate sensor data	55	sensitivity - privacy of data	41
interoperability	45	ensuring data quality	32
continuous measurements	41	costs of deployment and maintenance	32
open access to data	27	lack of interoperability	32
high temporal resolution	27	power-consumption	27
plug-and-play	27	missing metadata	27
scale-ability	23	undeveloped tools to discover and share data	27
well-developed tools to discover and share data	23	restricted data policies	23
spatially explicit measurements	23	data heterogeneity	18
high data quality	14	huge amount of data	18
content and context awareness*	5	funding in the long term	14
multi-sensor control*	5	technical knowledge needed in deploy and maintenance	14
decision support*	5	insecurity in communication	9
enable community participation	5	uncertainty of sensor performance	9
		pollution caused by abandoned sensors	9
		vandalism	5
		tools to use dynamically the data collected*	5

* respondents added items

4 Conclusions

The results of the expert survey on user requirements and future expectations for geosensor networks as described in this paper show that technology and applications are seen as main drivers for sensor network deployment. However harmonization of (open) standards to collect, access, manage, and integrate sensor data are considered a crucial boundary condition for further development. Although sensor network based data sources are increasingly used to provide real time data for everyday applications, their use in (autonomous) decision making requires further improvement of aspects like privacy and data quality. The respondents also see high potential in developing profitable business models from geosensor network applications in the future. Thus governmental funding is considered less important and data is seen as a valuable resource that is not always free for service providers and users. Finally, next to formal sensor networks, standardization will allow voluntary sensor information to become a significant sensor data source.

Acknowledgments. The authors acknowledge the financial support of the research program Space for Geo-Information (RGI) for organization of the workshop "Sensing a Changing World" in November 2008. We thank the participants of the workshop "Sensing a Changing World" for providing the answers to the survey.

References

1. Hart, J.K., Martinez, K.: Environmental Sensor Networks: A revolution in the earth system science? Earth Sci. Rev. 78, 177–191 (2006)
2. Nittel, S., Labrinidis, A., Stefanidis, A.: Introduction to Advances In Geosensor Networks. In: Nittel, S., Labrinidis, A., Stefanidis, A. (eds.) GSN 2006. LNCS, vol. 4540, pp. 1–6. Springer, Heidelberg (2008)
3. De Man, W.H.E.: Understanding SDI: Complexity and institutionalization. Int. J. Geogr. Inform. Science 20, 329–343 (2006)
4. Crompvoets, J., Bregt, A., Rajabifard, A., Williamson, I.: Assessing the worldwide developments of national spatial data clearinghouses. Int. J. Geogr. Inform. Science 18, 665–689 (2004)
5. Zhang, J., Hart, Q., Gertz, M., Rueda, C., Bergamini, J.: Sensor data dissemination using Webbased standards: a case study of publishing data in support of evapotranspiration models in California. Civ. Eng. Environ. Syst. 26, 35–52 (2009)
6. Delin, K.A.: The Sensor Web: A macro-instrument for coordinated sensing. Sensors 2, 270–285 (2002)
7. Botts, M., Percivall, G., Reed, C., Davidson, J.: OGC® Sensor Web Enablement: Overview and High Level Architecture. In: Nittel, S., Labrinidis, A., Stefanidis, A. (eds.) GSN 2006. LNCS, vol. 4540, pp. 175–190. Springer, Heidelberg (2008)
8. Le Cozannet, G., Hosford, S., Douglas, J., Serrano, J.J., Coraboeuf, D., Comte, J.: Connecting hazard analysts and risk managers to sensor information. Sensors 8, 3932–3937 (2008)
9. Kooistra, L., Bergsma, A., Chuma, B., de Bruin, S.: Development of a dynamic web mapping service for vegetation productivity using earth observation and in situ sensors in a sensor web based approach. Sensors 9, 2371–2388 (2009)

A Reference Architecture for Sensor Networks Integration and Management

Valentina Casola, Andrea Gaglione, and Antonino Mazzeo

Dipartimento di Informatica e Sistemistica
Universita' degli Studi di Napoli,
Federico II Naples, Italy
{casolav,andrea.gaglione,mazzeo}@unina.it

Abstract. Sensor networks have become a highly active research area due to their potential for providing diverse new capabilities for a wide variety of real world applications. Distributed applications require to collect information from a lot of different sensor systems, retrieved data are usually heterogeneous from many points of view and they need to be integrated to share a common objective. In this paper we present the SeNsIM framework, a scalable software architecture for the integration of heterogeneous sensor systems. SeNsIM enables the deployment of applications based on multiple sensor systems by providing a standard way to manage, query, and interact with sensors. We propose the architectural and data model of the SeNsIM framework and provide a method to describe sensor systems using XML as modeling language in order to facilitate sharing of structured data across them.

1 Introduction

Sensor networks have become a highly active research area due to their potential for providing diverse new capabilities for a wide variety of real world applications. They have made many novel applications possible to emerge in the fields of environmental monitoring [1], detection and classification of objects in military settings [2] and health applications [3]. Moreover, in the last years a proliferation of *Wireless Sensor Network (WSN)* technologies is increasing. Such systems are composed by low-cost and low-power sensor nodes (*"motes"*) able to measure different parameters and to communicate over wireless channels [9].

The diffusion of sensor systems, together with their applications have led to a large heterogeneity in the logic for interfacing and collecting data from these systems. As for WSNs, ad hoc programming languages (e.g. nesC [4]) and operating systems (e.g. TinyOS [20]) have been developed to support motes programming and to express the application processing in terms of message exchange among nearby nodes. Furthermore, different **middleware** platforms based on macro-programming models have been proposed in order to bridge the gap between the application and the underlying hardware and network platforms. However they are commonly used when a single application operates over a single WSN, while the application development for multiple WSNs is a rather cumbersome work.

N. Trigoni, A. Markham, and S. Nawaz (Eds.): GSN 2009, LNCS 5659, pp. 158–168, 2009.
© Springer-Verlag Berlin Heidelberg 2009

Nowadays the interaction with multiple sensor systems is required by a lot of applications, as those typical in the emerging pervasive computing paradigm adopted in several domains (e.g. telemedicine, crisis management, military [11]). Moreover monitoring applications of wide geographical areas have highlighted the research problem of the integrated management and correlation of data coming from various networks that cooperate for a common objective [6]. In order to embody sensing infrastructures into computing paradigms based on the application development for multiple WSNs or sensor systems, specific integration frameworks for accessing different data sources are needed.

This paper describes the design of the *SeNsIM (Sensor Networks Integration and Management)* framework, it is not just a middleware for WSNs, but it is a more general integration platform for heterogeneous sensor systems. SeNsIM (i) makes possible the deployment of applications based on multiple sensor systems/networks and (ii) allows a generic user or an application to easily access data sensed by a network. Furthermore, SeNsIM is able to ensure scalability since it is very easy to deploy new networks in the system.

The architectural model for the integration has been realized by exploiting a *wrapper-mediator* paradigm: the mediator accesses the sensor data by means of ad hoc connectors (wrappers), one for each network, which strongly depend on sensor network technologies. So the mediator is responsible to format and forward user requests to the different networks, while the wrappers are responsible to translate the incoming queries and forward them to the underlying sensors.

The remainder of this paper is organized as follows. In Section 2, an overview of solutions for WSN management is reported. In Section 3, we describe the proposed architectural model to manage sensor networks integration and the proposed data model for information exchange. In Section 4, some details on the system architecture and implementation are provided. Final discussions are reported in Section 5.

2   Related Work

In this section, the most well known **middleware** solutions for WSNs and some **integration platforms** for heterogeneous sensor systems are presented.

Middlewares for WSNs refers to software and tools which provide a system abstraction so that the application programmer can focus on the application logic without having to deal with the lower level implementation details [5]. Different middleware solutions for WSN management have been proposed [13,5,16]; they differ in terms of querying and data aggregation models and in the assumptions about the kind of sensor nodes, topology, size, and other features of the network. It is important to point out that all middlewares platforms are mainly focused on the possibility to access a single WSN implemented with a specific technology, but they do not provide the functionality to access different heterogeneous networks. To this end, some interesting research related to integration techniques for heterogeneous sensor networks has been done, but nowadays only a few architectures have been proposed.

In [8], Ahn and Chong propose an intelligent bridge as an interoperable architecture for message exchange between heterogeneous wireless sensor networks. They define a general message exchange mechanism that uses XML as message style and SOAP as transmission protocol. They also conducted a case study based on two WSNs with different network protocols (ZigBee and Bluetooth). Hourglass [18] provides an Internet-based infrastructure for connecting sensor networks to applications. It offers topic-based discovery and data-processing services, but focuses on maintaining quality of service of data streams in the presence of disconnections. GSNs (Global Sensor Networks) [7] facilitate the flexible integration and discovery of sensor networks and hide arbitrary stream data sources behind its virtual sensor abstraction. They enable the user to specify XML-based deployment descriptors in combination with the possibility to integrate sensor network data through plain SQL queries over local and remote sensor data sources. Other works try to define a middleware integration architecture that enables interoperability between sensor systems. For example the ESP framework [15] enables sensor systems to be queried without having to deal with the low-level implementation of specific access methods. It provides a mechanism to describe and model sensor systems using *ESPml*, an XML-based language, by which information regarding the sensor deployment can be specified. Finally, IrisNet [10] can be considered a hybrid approach to integration. It is a web infrastructure for easy deployment of wide-area sensing services. The architecture consists of *sensing agents* (SA) which collect and pre-process sensor data and of *organizing agents* (OA) which store sensor data in a hierarchical, distributed XML database that supports XPath queries.

Most of these approaches are still in a design phase since they lack a real implementation. Some try to define a common exchange mechanism among different sensor systems in order to facilitate the integration, but often they are strongly related to the adopted standards and technologies. Others are based on new technologies (i.e. web services for ESP), but the impact on performance is not discussed.

In contrast to the examined approaches, we will illustrate a more general architecture to provide a general-purpose infrastructure for sensor systems integration, and a more general way to express queries introducing a query visual language. Our system provides a complete data retrieval and management platform with a simple user interface by which it is possible to execute queries. Moreover our architecture is flexible and can easily provide APIs by which an application can easily interact with different sensor systems.

3   Design Principles

The need for deployment of applications based on multiple sensor systems and the management of wide geographical areas have highlighted a new research issue, i.e. the possibility of integrating and managing, in an integrated way, all data coming from the various data sources. Unfortunately sensed data are not available in a unique container, but are distributed in heterogeneous repositories. So the major issue for an integration system lies in the heterogeneity of repositories that makes data management and retrieval processes hard tasks to achieve. To face this problem, we need to define:

- an *architectural model* able to support in an efficient way the management of such data even when sensed by different networks;
- a *data model* capable of representing in a unique logical view the "sensor data", and that can be used by any application.

3.1   The Architectural and Data Model

As already said, sensor systems are adopted in many application fields and their proliferation has massively increased; due to these reasons there is a tremendous heterogeneity in the logic for interfacing and collecting data from these systems. In most sensor systems an application can directly access sensor hardware by means of opportune drivers. In the case of WSN, an operating system between the hardware platform and the application allows an easier use of the available sensing functions and often a further middleware layer between the operating system and the application provides an easy way to access sensor data. Furthermore sensed data may be structured differently according to the specific representation of different sensor systems.

We propose a novel integration platform (*SeNsIM, Sensor Networks Integration and Management*) which aims to bridge the gap between heterogeneous sensor systems and to provide a unique way to manage, query and interact with them. SeNsIM is made of a *mediator* component which hides networks heterogeneity to end users or applications by means of ad hoc connectors (*wrappers*) (one for each network). Each wrapper explores and monitors the local sensor network and sends to the mediator an appropriate description of the related information according to a common data model. The mediator organizes such information and keeps a unique view of all systems in order to satisfy user or application queries.

The architectural model we propose is made of four logical layers (see Fig. 1):

1. an application or user layer to submit queries and elaborate the retrieved data;
2. a mediator layer to format and forward queries to specific wrappers;
3. a wrapper layer to extract and manage network information and data;
4. the sensor system layer with or without a specific middleware or operating system.

As for the data model, in the literature, sensors have been modeled by using two kinds of approaches [17,19]: (i) structural approach focuses the attention on the sensor-structure in terms of hw/sw components, while (ii) data-oriented approach schematizes a sensor using a behavioral description. The latter mainly refers to sensor global information (type, producer, description, etc...) as well as to variables that a sensor can measure (temperature, light, humidity, etc...), predicates that a sensor can calculate (e.g. temperature greater than a threshold) and sensor operating state/mode (on, off, sleep, etc...).

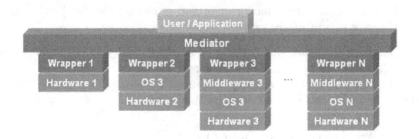

Fig. 1. Abstraction layers of the proposed architecture

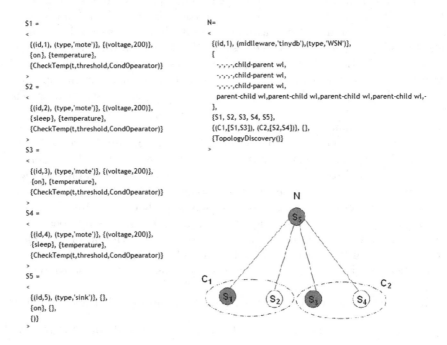

Fig. 2. An example of network description

In our approach we define a *sensor node* as an object characterized by a tuple of information that combines both structural and behavioral features. The state of a sensor can be modified by means of classical getting/setting functions, while the measured variables can be accessed using the sensing function. Collection of sensor nodes, placed according to clustering policies and to a given topology form a sensor network. According to our model a *network* object has to include global information such as type of sensor system, middleware (if present), supported sensor board as well as information related to sensor components (list of sensors, possible list of clusters, topology matrix). Network global predicates (e.g. average temperature of the network) also can be represented in our model.

For brevity, we do not report all the details of the data model, we just say that it is represented in XML format, since XML provides platform independence, interoperability and can be easily parsed [21]. XML-based descriptors have a unifying grammar by which systems can describe their abilities and define a standard language protocol with which the different entities in the framework can communicate. Our XML descriptor (*structXML*) is directly derived from the data model and represents features of both networks and sensors. Each wrapper builds a structXML descriptor after having injected a discovery query on the underlying system (generally at startup). If some parameters cannot be extracted in an automatic manner (i.e. sensor producer, middleware for WSN), a wrapper administrator can manually fill the structXML descriptor with the missing information.

As an example, Fig. 2 shows a network (N) and a pseudo-XML description of it. For simplicity we do not show the real structXML descriptor, trying to point out only the main features that could be of interest to the reader. The network is a WSN composed of 4 temperature sensor nodes (*S1* and *S3* are in the "on" state, while *S2* and *S4* are in the "sleep" state, being used as redundant units), grouped into two clusters (*C1* and *C2*), and of one *base station* (*S5*). The base station, also called *sink*, is a mote, in general connected with a PC-class device, which acts as a gateway between the network and the end user.

4   The Reference Architecture

In this section, we firstly describe the two main components of the SeNsIM system: the wrapper and the mediator. Then we illustrate interactions taking place between a generic wrapper, the mediator, the user and the underlying network during two main usage scenarios of SeNsIM.

As already said, wrapper components work as adapters between the mediator layer and the sensing platforms. They should gather the features of the underlying network and of its sensors (e.g. discovering the network topology with its clusters/groups of sensors, the state of single sensors, etc...) and, above all, they should be able to access sensor data by querying single sensors, clusters or the entire network. The mediator should classify sensor information sent by wrappers and should provide a simple way to users or applications for querying the networks. In Fig. 3 the architecture of wrapper and mediator components is reported, illustrating its main modules. The macro-modules of both components, which are represented in dashed lines, carry out the main features of the related component.

As for the wrapper, (i) it discovers, extracts and manages information about the underlying network and its sensors (*Network Classification*); (ii) it receives user queries from the mediator and executes them on the system by using its APIs and the local query language (*Query Processing*); (iii) it manages the communication process with the mediator (*Mediator Communication*). The Network Classification and Query Processing modules interact with a DBMS for storing and accessing information related to network/sensors (according to the data

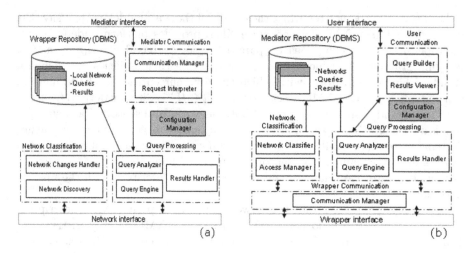

Fig. 3. (a) Wrapper architecture; (b) Mediator architecture

model), queries and related results. The wrapper is also provided with a *Configuration Manager* module, which the can be used by the system administrator to set the state of sensors or to define clustering/grouping policies during the discovery phase of the system.

The mediator, on the other side, (i) has to classify and manage metwork/sensor information sent by the wrapper (*Network Classification*); (ii) it has to manage user queries (*Query Processing*); (iii) it has to manage the communication with wrappers (*Wrapper Communication*); (iv) it has to interact with the user, by taking his queries and showing him the related results (*User Communication*). Also in this case the Network Classification and Query Processing modules interact with a DBMS which stores data related to networks and their sensors (intensional part of the data model), user queries and related results. Finally, the mediator is provided with a *Configuration Manager* module: it is used during the initialization phase of the system to define the admissible information for a network and its sensors according to the data model.

Fig. 4 shows the interactions that take place among the main system components in the two main usage scenarios, *registration* and *querying*:

– *Registration.* A wrapper needs to register itself before communicating with a mediator. At first, each wrapper creates an XML document describing the system as a whole; this is done by analyzing the sensor system (i.e. by injecting a discovery query), and generating the appropriate XML to represent the system. Then a wrapper sends a registration request message to the mediator, that verifies the possibility of including a new system in the framework and sends a response message to the wrapper. If the registration request is accepted, the wrapper sends the XML document to the mediator, which stores the related information in a DB.

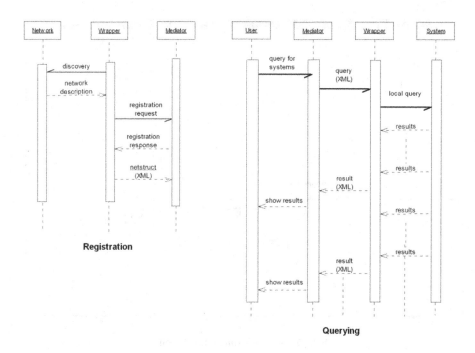

Fig. 4. The communication protocols in the Registration and Querying scenarios

– *Querying.* The querying process starts when a user sends a query request through the mediator user interface after having selected the destination of the query (a network or a specific sensor, among those connected to the framework trough the wrappers). The mediator takes the query parameters and creates an XML document, which is sent to the appropriate wrapper. The wrapper extracts the parameters by parsing the XML document and executes the query on the local system. The query results are grouped by the wrapper, which also creates another XML document, and periodically sends it to the mediator. Finally the mediator extracts the results from the XML and shows them to the user.

The system provides support for monitoring queries that retrieve the requested data from the sensor systems and return the corresponding responses in real-time as well as for event queries. Many context-aware applications need to trigger appropriate actions after that certain events have been detected by sensor systems.

4.1   Implementation Details

In the current implementation of SeNsIM, the mediator component and all its modules have been fully realized. Furthermore, we have implemented the wrapper component for TinyDB based networks. TinyDB [14] is a middleware for WSNs, which provides a query processing system for extracting information

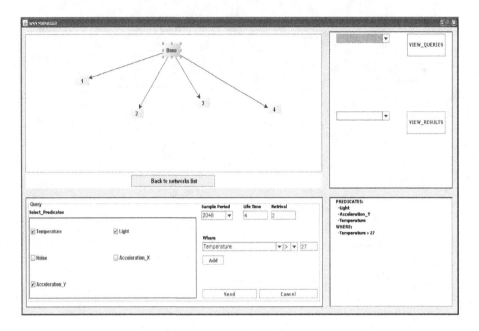

Fig. 5. Network querying screen

from a network of TinyOS sensors. Unlike existing solutions for data processing in TinyOS, TinyDB does not require users to write embedded C code for sensors. Instead, it provides a simple SQL-like interface to specify the data to extract, along with additional parameters, such as the data refreshing rate. Given a query, TinyDB collects that data from motes in the environment, filters it, aggregates it together, and routes it out to a PC, it also implements power-efficient in-network processing algorithms.

All SeNsIM components are written in the Java programming language exploiting its libraries for networking (java.net), multithreading (java.lang.Thread), managing DBMS (JDBC) and manipulating XML files (DOM). The communication protocol between wrappers and mediator has been implemented by exploiting both UDP and TCP sockets: in particular we used UDP sockets to exchange simple messages (e.g. registration request in the registration scenario), while TCP sockets were used to exchange XML data files (networks structure, queries and results). An ORACLE 10g DBMS has been adopted to store data in the wrapper and mediator repositories and a graphical interface has been realized to simplify the interaction of SeNsIM with users. Finally we set up a first experimental testbed with two TinyDB based networks in order to demonstrate the validity of our approach. Figure 5 provides a GUI screenshot showing the panel for querying a network after it has been selected by the user from the networks list.

5   Conclusions and Future Work

We have presented a system for the integration and management of sensor networks. It allows a single unified view of a lot of systems and enables a flexible deployment and interconnection between them, even if located in different places. The architecture is based on the Mediator/Wrapper paradigm in order to provide a layered and scalable architecture. The Wrapper is responsible for managing hardware and systems heterogeneity while the Mediator is responsible of managing the communication with the wrappers and provides a uniform interface for queries on sensed data and network features to user and applications. Future work will be devoted to evaluate system scalability and performance, and to develop service-oriented interfaces in order to integrate our system with recent standards proposed by the OGC and W3C consortiums for modeling sensors and sensor networks.

References

1. Steere, D., Baptista, A., McNamee, D., Pu, C., Walpole, J.: Research challenges in environmental observation and forecasting systems. In: Proc. ACM/IEEE MOBI-COM 2000, Boston (August 2000)
2. Hu, Y., Li, D., Wong, K., Sayeed, A.: Detection, classification, and tracking in distributed sensor networks. IEEE Signal Processing Magazine, 17–29 (March 2002)
3. Schwiebert, L., Gupta, S.K.S., Weinmann, J.: Research challenges in wireless networks of biomedical sensors. In: Proc. ACM/IEEE MOBICOM 2001, pp. 151–165 (2001)
4. Gay, D., Levis, P., Culler, D., Brewer, E.: nesC 1.1 Language Reference Manual (March 2003), http://nescc.sourceforge.net/papers/nesc-ref.pdf
5. Hadim, S., Mohamed, N.: Middleware for wireless sensor networks: A survey. In: Proc. 1st Int. Conf. Comm. System Software and Middleware (Comsware 2006), New Delhi, India, January 8-12 (2006)
6. Flammini, F., Gaglione, A., Mazzocca, N., Moscato, V., Pragliola, C.: Wireless Sensor Data Fusion for Critical Infrastructure Security. In: Proceedings of the International Workshop on Computational Intelligence in Security for Information Systems CISIS 2008 - Advances in Soft Computing, pp. 92–99 (2008)
7. Aberer, K., Hauswirth, M., Salehi, A.: The Global Sensor Networks middleware for efficient and flexible deployment and interconnection of sensor networks, Technical Report (2006)
8. Ahn, S., Chong, K.: Building a Bridge for Heterogeneous Sensor Networks. In: Proceedings of the Fourth IEEE Workshop on SEUS-WCCIA 2006 (2006)
9. Akyildiz, I.F., Vuran, M.C., Akan, O.B., Su, W.: Wireless Sensor Network: A survey REVISITED. Computer Networks Journal (2005)
10. Gibbons, P.B., Karp, B., Ke, Y., Nath, S., Seshan, S.: IrisNet: An Architecture for a World- Wide Sensor Web. IEEE Pervasive Computing 2(4) (2003)
11. Hadim, S., Mohamed, N.: Middleware: Middleware Challenges and Approaches for Wireless Sensor Networks. IEEE Distributed Systems Online (March 2006)
12. Heinzelman, W.B., Murphy, A.L., Carvalho, H.S., Perillo, M.A.: Middleware to support sensor network applications. IEEE Network 18(1), 6–14 (2004)

13. Henricksen, K., Robinson, R.: A Survey of Middleware for Sensor Networks: State-of-the-Art and Future Directions. In: MidSens 2006: Proceedings of the international workshop on Middleware for sensor networks, pp. 60–65. ACM Press, Melbourne (2006)
14. Madden, S., Franklin, M.J., Hellerstein, J.M., Hong, W.: TinyDB: An acquisitional query processing system for sensor networks. ACM Transactions on Database Systems 30(1), 122–173 (2005)
15. Reddy, S., Schmid, T., Yau, N., Chen, G., Estrin, D., Hansen, M., Srivastava, M.B.: ESP Framework: A middleware architecture for heterogeneous sensing systems (December 2006)
16. Romer, K.: Programming Paradigms and Middleware for Sensor Networks, GI/ITG Fachgespraech Sensornetze, Karlsruhe, Feburary 26-27 (2004)
17. SensorML Project, http://vast.uah.edu/SensorML/
18. Shneidman, J., Pietzuch, P., Ledlie, J., Roussopoulos, M., Seltzer, M., Welsh, M.: Hourglass: An Infrastructure for Connecting Sensor Networks and Applications, Technical Report TR-21-04, Harvard University, EECS (2004), http://www.eecs.harvard.edu/Usyrah/hourglass/papers/tr2104.pdf
19. Skov, T., Bro, R.: A new approach for modelling sensor based data. Sensor and Actuators B: Chemical 106(2), 719–729 (2005)
20. TinyOS Project, http://www.tinyos.net
21. W3C Architecture Domain, Extensible Markup Language (XML), http://www.w3.org/XML/

A Stimulus-Centric Algebraic Approach to Sensors and Observations

Christoph Stasch[1], Krzysztof Janowicz[1], Arne Bröring[1],
Ilka Reis[2], and Werner Kuhn[1]

[1] Institute for Geoinformatics, University of Muenster, Germany
{staschc,janowicz,arneb,kuhn}@uni-muenster.de
[2] Departamento de Estatística
UFMG and National Institute for Space Research (INPE), Brazil
ilka@dpi.inpe.br

Abstract. The understanding of complex environmental phenomena, such as deforestation and epidemics, requires observations at multiple scales. This scale dependency is not handled well by today's rather technical sensor definitions. Geosensor networks are normally defined as distributed ad-hoc wireless networks of computing platforms serving to monitor phenomena in geographic space. Such definitions also do not admit animals as sensors. Consequently, they exclude human sensors, which are the key to volunteered geographic information, and they fail to support connections between phenomena observed at multiple scales. We propose definitions of sensors as information sources at multiple aggregation levels, relating physical stimuli to observations. An algebraic formalization shows their behavior as well as their aggregations and generalizations. It is intended as a basis for defining consistent application programming interfaces to sense the environment at multiple scales of observations and with different types of sensors.

1 Introduction

Sensor networks have become an important technology for observing physical phenomena. Their application scenarios include environmental and health monitoring, disaster management, early warning systems, precision agriculture, as well as home security [1]. Currently, most research is focused on technical issues. This includes work on hardware, operation systems, and signal processing [2], but also algorithmic aspects about how to reduce power consumption [3] and communication costs [4]. In contrast, this paper investigates sensors from an ontological perspective.

The understanding of complex phenomena, such as deforestation and epidemics, requires environmental sensor networks [5] observing at multiple scales. This scale dependency is not handled well by today's rather technical sensor definitions and therefore also in the derived models. For instance, to understand the impact of deforestation on the local fauna, it is necessary to track the path of individuals as well as the path of populations within a biotope. Movement patterns of individuals reveal information about change in territory and foraging,

N. Trigoni, A. Markham, and S. Nawaz (Eds.): GSN 2009, LNCS 5659, pp. 169–179, 2009.

while the changed behavior of one population impacts the behavior of others. At the scale of the population, a sensor network should produce a single trajectory based on the tracks of the individual animals. Current definitions of sensors, sensor systems, and sensor networks are too technical to capture these abstractions of observations. For example, the definition of geosensor networks as 'distributed ad-hoc wireless networks of sensor-enabled miniature computing platforms that monitors phenomena in geographic space' [6] does not admit animals as sensors. It restricts the notion of sensors and sensor networks to technical devices and the message transmission to wireless communication.

In our work, we extend the second part of Nittel's definition by focussing on the scale dependency of observation regarding various sensor aggregation levels. Our main contribution is an algebraic specification of sensors. It provides a unified view on sensors at different scales, i.e., at different spatial, temporal and thematic granularity of observed phenomena. Our approach bridges between the sensor-centric Sensor Model Language (SensorML) [7] of the Open Geospatial Consortium (OGC) and the user-centric Observations & Measurements (O&M) [8] specification by focusing on stimuli as objects of sensing. We also include human sensors which are the key to volunteered geographic information [9].

The remaining paper is structured as follows. First, we give an overview of related work. We then propose to define sensors as information sources at multiple scales, relating physical stimuli to observations. An algebraic formalization shows their aggregations, compositions, and generalizations. We provide various examples from real applications of sensors and sensor systems to demonstrate our approach. The paper closes with conclusions and an outlook to future work.

2   Related Work

Current sensor models and definitions are designed from a technical perspective. In the engineering community, sensors are defined as devices that produce analog signals based on the observed phenomenon. These signals are converted to digital signals by analog-to-digital converters (ADCs). Sensor networks comprise a large number of sensor nodes 'that are densely deployed either inside the phenomenon or very close to it' [10]. A taxonomy of distributed sensor networks based upon different criteria such as input or communication could be found in [11]. Sgroi et al. [12] developed a set of well defined services and interface primitives for programming of ad-hoc wireless sensors and actuator networks. The SensorML specification [7] defines a common model for all kinds of sensor related processes, whereas a sensor is defined as a process entity capable of observing a phenomenon and returning an observed value. Our work is similar to [12,7] but is focused on the ontological view on sensors. Additionally, we provide a unique ontological view on sensors on different scales and include humans and animals as sensors.

There are several topics of current research on sensor networks. A lot of work has been done on reducing the in-network communication cost to reduce energy consumption by reducing the amount of data [4,13,14] or optimizing the data collection path [15]. As GSNs usually monitor dynamic phenomena in space, in-network detection of changes or events regarding the observed phenomenon is also investigated [16]. Another issue currently addressed is the localization of sensors in geosensor networks [17]. As the amount of data has to be reduced for further processing, several abstraction mechanisms for the sensor observations are presented [18]. Recently, Goodchild [9] proposed to extend geosensor networks to include humans either as sensor platforms or as sensors themselves. These human sensor networks could serve as the basis for the Volunteered Geographic Information enabled by Web 2.0 technologies. An example in this context is the bird post application (www.birdpost.com) which enables its users to report bird sightings or to search for bird sightings by location or characteristics.

To enable the web based exchange of geosensor data and the integration of sensor data into spatial data infrastructures, OGC's Sensor Web Enablement (SWE) initiative provides a framework of standards for the realization of a Sensor Web. Following Botts et al. [19] a Sensor Web refers to web accessible sensor networks and archived sensor data that can be discovered and accessed using standard protocols and application programming interfaces (APIs). Research on the Semantic Sensor Web [20] investigates the role of semantic annotation, ontologies, and reasoning to improve discovery on the Sensor Web [21]. It combines OGC's vision of a web of sensors with the reasoning capabilities of the Semantic Web. Besides discovery, a semantic layer would improve interoperability between sensor networks and would help to make sensors situation aware. A method for linking geosensor databases with ontologies has been presented by Hornsby [22]. An ontological analysis of the OGC standards on observations and measurements has been done by Probst [23]. However, the integration of semantics into sensor networks and sensor applications is still a challenging research task and a thoroughly defined model for sensors from an ontological perspective is currently missing.

3   An Algebraic Approach to the Aggregation of Sensors

In this section we introduce algebraic specifications for sensors, sensor systems, and sensor networks as well as for their aggregation and generalization. For lack of space and to improve readability we focus on particular aspects, leaving others (especially details about location, time, and signal processing) aside. The evolving, executable source code, specified using the functional language Haskell[1], is available online at http://musil.uni-muenster.de/gsn09hs.

3.1   Stimuli and Their Observation

Our knowledge about the real world is based on observations. This includes observations about endurants such as a single zebra [24] or perdurants (also called occurants [25]) such as the dispersion of diseases like the dengue fever. However, in most cases what is observed are not such particulars themselves but

---

[1] Introductions to the Haskell language, as well as interpreters and compilers can be found at http://www.haskell.org

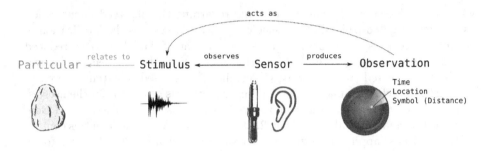

Fig. 1. Particulars such as a stone (hitting a water surface) can only be observed by stimuli which are related to them. The sensor has no knowledge about the particular itself (indicated by the greyed out color). Stimuli can only be defined with respect to a specific sensor. For example, the sound waves from the impact of the stone hitting the water surface are stimuli for human ears or a (passive) sonar. Stimuli are located and timed. For instance, the impact produces sound waves for a certain amount of time spreading from the location where the stone hits the surface in radial direction. While the stimulus acts as input to the sensor, the output is an observation. Such observations can be stimuli for other sensors (as in SensorML [7], though stimuli and observations are not not distinct types in SensorML), e.g., to produce a terrain model.

physical processes [26] related to them. We call those physical processes *stimuli* for which we have sensors. For instance sound waves within a certain frequency range become stimuli when they reach the human ear; they reveal something about particulars in the real world. We learn about particulars like walls from their echoing of sound waves. In this case the stimulus is created by the sensor itself, e.g., by an active sonar. In case of passive sensors such as human ears or

Listing 1. Stimuli as observable spatio-temporal referenced processes

```
0    class LOCATED item where
1           getLocation :: item -> Location
2
3    class TIMED item   where
4           getTime :: item -> Time
5
6    data Process = SoundWave | EggReflection | Signal Observation |
7
8    data Stimulus = Stimulus Process Location Time
9
10   instance LOCATED Stimulus where
11          getLocation (Stimulus p l t) = l
12
13   instance TIMED Stimulus where
14          getTime (Stimulus p l t) = t
15
16   eggs = Stimulus EggReflection  'Porto Seguro' 04/2009
17   sound = Stimulus SoundWave theSonarsLocation now
```

Listing 2. Observations as sensor results

```
18   data Observation = Observe [Stimulus]
19
20   instance LOCATED Observation where
21           getLocation (Observe stimuli) =
22                           getLocation (last stimuli)
23
24   instance TIMED Observation where
25           getTime (Observe stimuli) = getTime (last stimuli)
26
27   eggCount = Observe [eggs]
28   echoImpact = Observe [sound]
29
30   countEggs = observe aHumansVision [eggs]
31   receiveEcho = observe aSonar [sound]
```

passive sonars, the stimulus are the sound waves caused by a certain event (see figure 1). Another example is the problem of monitoring dengue fever. To make predictions about the future spread of this disease we observe the presence of eggs of *aedes aegypti* [27], the transmitters of dengue fever. We have several sensors to make this observation. For instance, we can detect the reflected light (stimulus) using our eyes (sensor) or the resistance from the solid surface (stimulus) with our sense of touch (sensor). In this case, a perdurant (the spread of a virus disease) is forecasted by observing another perdurant, namely the presence of eggs of the transmitting mosquito in a given region.

Note that we only require minimal assumptions about reality here. All information about reality is derived from observations and depends on sensors (including the human sensory system). In contrast to the user-centric view taken in OGC's O&M specification [8], our view focuses on stimuli as objects of sensing. At the same time, it abstracts from the technical, sensor-centric view of the SensorML specification [7]. In this way, our view bridges between SensorML and O&M [8].

Line 16 shows one specific stimulus, the occurrence of eggs of *aedes aegypti* in the Brazil city Porto Seguro in April 2009 (detected by the reflection of the eggs). This example also shows that we constrain the temporal and spatial resolution by the spatio-temporal extent of the stimulus. The behavior of stimuli is defined by the data type STIMULUS specified in line 8-14. We model this extent of stimuli by introducing the type classes TIMED and LOCATED (line 0-4) which provides location and time on an abstract level.

We use a broader definition of OBSERVATION here than the one proposed by OGC [8]. Observations, as sensor results, are spatio-temporally referenced symbols provided to a user or for further processing (line 18-25). Symbols are observable signals and hence can be perceived by sensors again. Such symbols range from a count such as an integer for the number of eggs to complex symbols such as *mostly cloudy with occasional rain* in case of weather conditions. In the first case, an egg count is an observation where the human vision acts as sensor (line 27 & 30). In the second case, the literal *mostly cloudy with occasional rain* produced by the weather station is perceived by the human eyes acting as sensors (see figure 1). The result time and location of the observation may differ from those of the stimulus. Their georeferencing is up to the sensor (see below). Note that observations are processes (line 6) and can act as stimuli for other kinds of sensors.

3.2 Sensors

The SENSORS type class (line 32-34) defines the behavior of kinds of sensors by providing the *observe* function (line 33). It maps from stimuli to observations. An example for a sensor is a heart frequency sensor attached to an animal. In this case, the location is not georeferenced and of minor interest. The implementation of the observe function is up to the specific kind of sensor. It captures how the stimulus is transformed to a digital signal, processed, and finally mapped to the symbol. This behavior should be described using standards such as SensorML. GEOSENSORS are kinds of sensors delivering an observation with georeferenced location (line 38). An example is the EGGSENSOR type for counting the eggs of *aedes aegypti*. The observations of the EggSensor contain a georeferenced location. Sensors are mounted on platforms (line 41) which provide an ID and a tracking function for their location. For instance, human eyes are placed on the human body, while a multi-spectral camera is mounted on a satellite.

Listing 3. Sensors mapping stimuli to observations

```
32   class LOCATED sensor => SENSORS sensor where
33          observe :: sensor -> [Stimulus] -> Observation
34          observe sensor = Observe
35
36   data Sensor = Sensor Platform Process
37
38   data Location = GeoLocation Coordinates Epsg |
39                   BioLocation Organ Id | ...
40
41   data Platform = Platform Id (ClockTime -> Location)
```

Aggregation. Sensors can be aggregated to sensor systems or sensor networks by using the *addSensor* operation (line 43). The *getSensor* operation (line 44) can be used to access a single sensor interface of the sensor aggregation to retrieve its observations.

A SENSORSYSTEM aggregates sensors (line 49-59) which are mounted on the same platform. An example is an aggregation of heart frequency and blood pressure sensors to provide the overall cardiac condition of an animal.

A GEOSENSORSYSTEM contains at least one geosensor defining the georeferenced locations of its observations. An example is a weather station (line 65,66) consisting of a thermometer and a barometer which are all georeferenced.

Listing 4. Sensors systems as sensor aggregations

```
42  class SENSORSYSTEMS sensorsystem where
43      addSensor :: Sensor -> sensorsystem -> sensorsystem
44      getSensors :: sensorsystem -> [Sensor]
45
46  data SensorSystem = SensorSystem Platform [Sensor]
47  aHumansSenses = SensorSystem aHuman [Hearing, Vision,...]
48
49  instance SENSORSYSTEMS SensorSystem where
50      addSensor (Sensor platform2 stimulus)
51      (SensorSystem platform1 stimuli) =
52          if platform1 == platform2 then SensorSystem platform1
53          (stimulus : stimuli)
54          else error "incompatible_platform"
55      getSensors (SensorSystem platform [stimulus]) =
56          [Sensor platform stimulus]
57      getSensors (SensorSystem platform stimuli) =
58          (Sensor platform (head stimuli)) :
59              getSensors (SensorSystem platform (tail stimuli))
60
61  instance LOCATED SensorSystem where
62      getLocation (SensorSystem platform stimuli) time =
63          getLocation platform time
64
65  aWeatherStation = SensorSystem aPlatform
66                      [aThermometer, aBarometer, ...]
```

A SENSORNETWORK consists of a number of spatially distributed and communicating sensors or sensor systems. Instead of specific communication paradigms we only state that one can request the neighbors of a specific sensor. In an implementation of this specification the *getNeighbors* operation (line 71) could be realized by delegating to a corresponding operation of the sensor nodes. An example of a sensor network is a spatially distributed and connected group of animals carrying sensor systems of heart frequency and blood pressure sensors.

A GEOSENSORNETWORK is a sensor network whose nodes are all geosensors. For example, in the OSIRIS project [28], air quality sensors have been deployed on moving buses. These sensors taken together form a network for urban air quality monitoring, as they are spatially distributed and connected through wireless communication.

If a sensor system or a sensor network is a composition of sensors, all subsensors of the system or the network can only exist in it. One example for a composition of sensors is the human sensor system. The human eye cannot exist without the human body it belongs to.

Generalization. Both, sensor systems and sensor networks can behave like single sensors by providing the observe function. Therefore, the proposed sensor aggregation realizes the *compositum* design pattern [29]. Thus, on different the-

Listing 5. Sensors networks as sensor aggregations

```
67   class SENSORS sensor ⇒ SENSORNETWORKS sensornetwork sensor where
68        addSensor :: sensor -> sensornetwork sensor
69                   -> sensornetwork sensor
70        getSensors :: sensornetwork sensor -> [sensor]
71        getNeighbors :: sensor -> sensornetwork sensor -> [sensor]
```

matic, temporal or spatial scales a sensor aggregation can be generalized to a single sensor. The resulting symbol could be either a complex symbol containing single sensor symbols or a generalized symbol. In the example of the air quality monitoring network, if the sensor network is generalized to a single sensor, it either provides a value grid, a generalized value (e.g. an average) or a nominal value (e.g. 'smog'). The generalization defined by our sensor algebra affects the spatial and temporal resolution of the observation as well as the symbol resolution (as shown in the air quality example).

4   Conclusions and Further Work

In this paper we introduced an ontological definition of sensors at various scales. We provided algebraic specifications to show the aggregation, composition, and generalization of sensors. Our formalization is less restrictive in comparison to existing standards, and is more flexible in handling the heterogeneity of information sources. In contrast to more technical definitions, our approach explicitly includes human (and animal) sensors which play an important role in volunteered geographic information and environmental modeling.

A sensor implements a mapping from a physical stimulus to an observation. In contrast to the user-centric view taken in OGC's O&M specification and to the sensor-centric view in OGC's SensorML specification, our approach is stimulus-centric. It attaches the primary spatio-temporal context to the stimulus instead of the sensor. Geosensors differ from other sensors by providing additional georeferencing functionalities. This only requires minimal assumptions about reality, i.e., all information about reality is derived from observations and hence depends on sensors. Consequently, we do not need to make assumptions about objects (features of interest) at the sensor level. Observations can be stimuli for other sensors to enable more complex sensor processing chains. Sensors can be combined into sensor systems and networks. Depending on the thematic scale, these in turn can behave like a single sensor or a (spatially distributed) collection of sensors. The symbols of the aggregated sensors can be combined and generalized to create more complex symbols.

Further work will focus on defining more detailed algebraic specifications for the aggregation and generalization of sensors to reason about the resulting symbols and symbol systems. While geosensors rely on spatial reference systems to locate an observation, semantic reference systems [30] are necessary to reference complex symbols used by the sensor [31]. Such referencing, called *grounding*, is also required for the measurement procedure to fix the way how stimuli are mapped to observations.

The algebraic specifications developed in the paper will serve as a basis for defining a consistent application programming interface (API) to sense the environment at multiple scales of observations and with different devices. This involves work on defining a taxonomy of sensor types and relations between them. The API will allow for abstracting from specific sensor interfaces. It will focus on types of sensors and kinds of observations – making it a meta API.

Acknowledgments

The comments from three anonymous reviewers provided useful suggestions to improve the content and clarity of the paper. Major parts of the work presented here were developed at the Research Workshop of the Institute for Geoinformatics, University of Münster, Germany, and the National Institute for Space Research (INPE), Brazil, (see http://geochange.info), which has been supported by the German Research Foundation (DFG), project no. 444 BRA 121/2/09 and by The State of São Paulo Research (FAPESP), project no. 2008/11604-6.

References

1. Shepherd, D., Kumar, S.: Microsensor Applications. In: Distributed Sensor Networks. Chapman & Hall, Boca Raton (2005)
2. Iyengar, S., Brooks, R. (eds.): Distributed Sensor Networks, 1st edn., December 2005. Chapman & Hall / Crc Computer and Information Science (2005)
3. Markham, A., Wilkinson, A.: A Biomimetic Ranking System for Energy Constrained Mobile Wireless Sensor Networks. In: Southern African Telecommunications, Networks and Applications Conference (SATNAC), Mauritius, September 9-13 (2007)
4. Reis, I.A., Camara, G., Assuncao, R., Monteiro, A.: Suppressing Temporal Data in Sensor Networks using a Scheme Robust to Aberrant Readings. International Journal of Distributed Sensor Networks (to be published, 2009)
5. Hart, J., Martinez, K.: Environmental Sensor Networks: A revolution in the earth system science? Earth-Science Reviews 78, 177–191 (2006)
6. Nittel, S., Stefanidis, A.: GeoSensor Networks and Virtual GeoReality. In: GeoSensors Networks. CRC, Boca Raton (2005)
7. Botts, M.: OpenGIS Sensor Model Language (SensorML) Implementation Specification (2007)
8. Cox, S.: Observations and Measurements - Part 1 - Observation schema (2007)
9. Goodchild, M.: Citizens as Sensors: the World of Volunteered Geography. GeoJournal 69(4), 211–221 (2007)
10. Akyildiz, I., Tommaso, M., Kaushik, C.: A Survey on Wireless Multimedia Sensor Networks. Computer Networks 51(4), 921–960 (2007)
11. Sastry, S., Iyengar, S.: A Taxonomy of Distributed Sensor Networks. In: Distributed Sensor Networks, pp. 29–43. Routledge (2004)

12. Sgroi, M., Wolisz, A., Sangiovanni-Vincentelli, A., Rabaey, J.: A Service-Based Universal Application Interface for Ad Hoc Wireless Sensor and Actuator Networks. In: Weber, W., Rabaey, J., Aarts, E. (eds.) Ambient intelligence. Springer, Heidelberg (2005)
13. Akkaya, K., Younis, M.: A Survey on Routing Protocols for Wireless Sensor Networks. Ad Hoc Networks 3(3), 325–349 (2005)
14. Lin, S., Kalogeraki, V., Gunopulos, D., Lonardi, S.: Efficient Information Compression in Sensor Networks. International Journal of Sensor Networks 1(3/4), 229–240 (2006)
15. Kulik, L., Tanin, E., Umer, M.: Efficient Data Collection and Selective Queries in Sensor Networks. In: Nittel, S., Labrinidis, A., Stefanidis, A. (eds.) GSN 2006. LNCS, vol. 4540, pp. 25–44. Springer, Heidelberg (2008)
16. Worboys, M., Duckham, M.: Monitoring Qualitative Spatiotemporal Change for Geosensor Networks. International Journal of Geographical Information Science 20, 1087–1108 (2006)
17. Reichenbach, F., Born, A., Nash, E., Strehlow, C., Timmermann, D., Bill, R.: Improving Localization in Geosensor Networks through Use of Sensor Measurement Data. In: Cova, T.J., Miller, H.J., Beard, K., Frank, A.U., Goodchild, M.F. (eds.) GIScience 2008. LNCS, vol. 5266, pp. 261–273. Springer, Heidelberg (2008)
18. Jung, Y., Nittel, S.: Geosensor Data Abstraction for Environmental Monitoring Application. In: Cova, T.J., Miller, H.J., Beard, K., Frank, A.U., Goodchild, M.F. (eds.) GIScience 2008. LNCS, vol. 5266, pp. 168–180. Springer, Heidelberg (2008)
19. Botts, M., Percivall, G., Reed, C., Davidson, J.: OGC Sensor Web Enablement: Overview And High Level Architecture. Technical report, Open Geospatial Consortium (2007)
20. Sheth, A., Henson, C., Sahoo, S.: Semantic Sensor Web. IEEE Internet Computing, 78–83 (2008)
21. Jirka, S., Bröring, A., Stasch, C.: Discovery Mechanisms for the Sensor Web. Sensors 9 (in press, 2009)
22. Hornsby, K., King, K.: Linking geosensor network data and ontologies to support transportation modeling. In: Nittel, S., Labrinidis, A., Stefanidis, A. (eds.) GSN 2006. LNCS, vol. 4540, pp. 191–209. Springer, Heidelberg (2008)
23. Probst, F.: Ontological Analysis of Observations and Measurements. In: Raubal, M., Miller, H.J., Frank, A.U., Goodchild, M.F. (eds.) GIScience 2006. LNCS, vol. 4197, pp. 304–320. Springer, Heidelberg (2006)
24. Zhang, P., Sadler, C., Lyon, S., Martonosi, M.: Hardware Design Experiences in ZebraNet. In: 2nd international conference on Embedded networked sensor systems, pp. 227–238. ACM, New York (2004)
25. Masolo, C., Borgo, S., Gangemi, A., Guarino, N., Oltramari, A.: WonderWeb deliverable D18 ontology library. Technical report, IST Project 2001-33052 WonderWeb: Ontology Infrastructure for the Semantic Web (2003)
26. Hansman, J.: Characteristics of Instrumentation. In: Measurement, Instrumentation, And Sensors Handbook. CRC Press LLC, Boca Raton (1999)
27. Regis, L., Monteiro, A., Melo-Santos, M., Silveira, J., Furtado, A., Acioli, R., Santos, G., Nakazawa, M., Carvalho, M., Ribeiro, P., Souza, W.: Developing New Approaches for Detecting and Preventing Aedes Aegypti Population Outbreaks: Basis for Surveillance, Alert and Control System. Memorias do Instituto Oswaldo Cruz 103, 50–59 (2008)
28. Jirka, S., Bröring, A., Stasch, C.: Applying OGC Sensor Web Enablement to Risk Monitoring and Disaster Management. In: GSDI 11 World Conference, Rotterdam, Netherlands (June 2009)

29. Gamma, E., Helm, R., Johnson, R., Vlissides, J.: Design Patterns: Elements of Resusable Object-Oriented Software. Addison-Wesley Professional, Reading (1995)
30. Kuhn, W.: Semantic Reference Systems. International Journal of Geographic Information Science 17(5), 405–409 (2003)
31. Scheider, S., Janowicz, K., Kuhn, W.: Grounding geographic categories in the meaningful environment,
    `http://musil.uni-muenster.de/wp-content/uploads/2009/04/`
    `groundingcategoriesinme_final.pdf`(forthcoming, 2009)

Author Index